Electricity in buildings

CIBSE Guide K

CIBSE

Typeset by CIBSE Publications

Printed in Great Britain by Page Bros. (Norwich) Ltd., Norwich, Norfolk NR6 6SA

Note from the publisher

This publication is primarily intended to provide guidance to those responsible for the design, installation, commissioning, operation and maintenance of building services. It is not intended to be exhaustive or definitive and it will be necessary for users of the guidance given to exercise their own professional judgement when deciding whether to abide by or depart from it.

Foreword

The publication of CIBSE Guide K: *Electricity in buildings* fills a significant gap in the range of publications produced by the CIBSE.

Work on the Guide has been carried out intermittently for more than 14 years. I wish to express my thanks to all those who have contributed their time and knowledge both under, and prior to, my Chairmanship.

Guide K is not intended as a textbook for those wishing to learn electrical engineering but rather as a reference source for engineers, architects, surveyors and other professionals who requires easy access to the information contained therein. Similarly, it is not intended that this Guide should be read from beginning to end in one go. Instead, the reader should dip into the relevant section that is of particular interest in respect of the project in hand.

Whilst the authors and contributors have tried to produce a document that will prove to be a valuable source of reference and guidance, it is inevitable that some users will find areas that they consider should include further information. Constructive criticism and comment is welcomed in order to improve and extend the document.

Finally I wish to thank the staff of the CIBSE Publications Department and the members of the Guide K Steering Committee for their valuable assistance in the publication of the Guide.

Richard Cleaver
Chairman, CIBSE Guide K Steering Committee

Guide K Task Group

Richard Cleaver (Chairman) (Ridge)
Ian Bitterlin (Emerson Network Power Ltd)
Roger Cunnings (Zisman Bowyer & Partners)
Jeff Dobson (G K Salter & Associates)
Alwyn Finney (ERA Technology Ltd)
Mick Geeson (N G Bailey & Co Ltd)
Ian Harrison (H I Associates & Co)
John Hewitt (Arup R+D)
Kevin Kelly (ABB Durham Switchgear Ltd)
Anthony Lo Pinto (Jaros Baum & Bolles)
Howard Porter (Arup R+D)
Roy Waknell (Power Systems Design Ltd)

Principal authors and contributors

Section 1: Introduction

Principal author
Richard Cleaver (Ridge)

Section 2: Legislation and standards

Principal authors
John Hewitt (Arup R+D)

Section 3: Load assessment

Principal author
Anthony Lo Pinto (Jaros Baum & Bolles)

Section 4: Power supplies

Principal authors
Richard Cleaver (Ridge)
Roger Cunnings (Zisman Bowyer & Partners)

Section 5: High voltage switchgear and distribution

Principal authors
Ian Bitterlin (Emerson Network Power Ltd)
Richard Cleaver (Ridge)

Section 6: Transformers

Principal authors
Kevin Kelly (ABB Durham Switchgear Ltd)
Richard Cleaver (Ridge)

Section 7: Low voltage switchgear and distribution

Principal author
Ian Pegrum (Faber Maunsell plc)

Section 8: Building wiring systems

Principal author
Howard Porter (Arup R+D)

Section 9: Uninterruptible power supplies

Principal authors
Ian Bitterlin (Emerson Network Power Ltd)
Jeff Dobson (G K Salter & Associates)
Ian Harrison (H I Associates & Co)

Section 10: Earthing

Principal author
Roy Waknell (Power Systems Design Ltd)

Section 11: Electromagnetic compatibility

Principal author
Alwyn Finney (ERA Technology Ltd)

Section 12: Inspection, testing, operation and maintenance

Principal author
Mick Geeson (N G Bailey & Co Ltd)

Editor

Ken Butcher

CIBSE Project Manager

Alan C Watson

CIBSE Publishing Manager

Jacqueline Balian

Acknowledgement

Crown copyright material is reproduced with the permission of the Controller of HMSO and the Queen's Printer for Scotland under licence number C02W0002935.

Contents

1 Introduction

1.1 General

The CIBSE has a large number of members who are involved in electrical services but its only publication covering the mainstream of electrical building services has been CIBSE Guide B10, withdrawn some years ago.

CIBSE Guide K: *Electricity in buildings* has been produced to fill the gap for electrical knowledge related to building services systems.

1.2 Purpose of the Guide

It is intended that this Guide will be used in conjunction with established codes to provide guidance to practitioners. It will also be of interest to designers and authorities who, whilst not directly concerned with the design or installation of electrical systems, need to understand the advice offered to them by specialists. Furthermore, the Guide should be of value to students embarking on a career in electrical or building services engineering and those already practising in these disciplines who wish to enhance their knowledge through a programmed of continuing professional development.

1.3 Contents of the Guide

1.3.1 Section 1: Introduction

Section 1 discusses the purpose and scope of the Guide and provides an overview of its structure and contents.

1.3.2 Section 2: Legislation and standards

This section defines the current legislation relating to electrical services and identifies the British and European Standards that are used in the design and installation of electrical installations.

1.3.3 Section 3: Load assessment

This section considers methods of assessing electrical loads within buildings as a prerequisite to carrying out design and to defining the size of the incoming supply. This section also gives consideration to the effect of load assessment on the supply arrangements with regional electricity companies.

1.3.4 Section 4: Power supplies

Section 4 discusses the external power supplies provided by the regional electricity companies, both at high voltage and low voltage. The section also discusses alternative sources of supply.

1.3.5 Section 5: High voltage switchgear and distribution

This section examines different types of circuits and the available types of high voltage switchgear, and discusses the options for the selection and erection of different types of equipment. The section discusses types of cabling and fault level protection in electrical systems.

1.3.6 Section 6: Transformers

Section 6 examines different types of power transformers and discusses the options available for the selection and erection of equipment.

1.3.7 Section 7: Low voltage switchgear and distribution

Section 7 examines the various types of low voltage switchgear and the forms of switchgear construction. The section also considers protection, metering and substation arrangements.

1.3.8 Section 8: Building wiring systems

This section examines the various types of wiring systems in use for final wiring circuits and other electrical applications. The section discusses the construction and use of various types of cables. Part of the section is devoted to considering the opportunities for prefabrication of wiring systems.

1.3.9 Section 9: Uninterruptible power supplies

This section describes the different types of UPS systems available and gives guidance on the selection of the most appropriate equipment for the required application. The section also includes a consideration of the different types of batteries available for use in UPS systems.

1.3.10   Section 10: Earthing and bonding

Earthing and bonding is an essential part of any electrical system and this section considers the requirements for the selection and erection of equipment to ensure electrical safety within buildings. The section includes discussion on the provision of lightning protection to buildings.

1.3.11   Section 11: Electromagnetic compatibility

This section examines various forms of electrical interference within the working environment and discusses methods for minimising the effects of this.

This section examines the need for electromagnetic compatibility and defines the various types of electrical interference that can occur. The section provides guidance on the elimination of electrical interference within buildings.

1.3.12   Section 12: Inspection, testing, operation and maintenance

Many construction projects involve the remodelling or refurbishment of buildings without the extensive installation of new services. This section considers the issues involved in the inspection and testing of existing installations both prior to alteration and as a periodic maintenance function during the life of the building. The section also considers the day-to-day operation of electrical equipment and the maintenance required to ensure safe operation.

The inspection, testing and verification of an installation is essential in confirming both its safety and operation.

Section 12 looks at the initial verification of a new installation and the requirements for periodic inspection of existing installations. Reference is made to both BS 7671[1] and IEE Guidance Note 3: *Inspection and testing*[2].

In carrying out either of these activities, the correct use of certification is necessary to comply with BS 7671 and, in doing so, providing evidence for the safety of an installation. This section details types of certification and reports, and the contents of those documents.

Section 12 also considers the need for operation and maintenance (O&M) manuals and the prescribed contents of such manuals, together with reference to types of maintenance and maintenance procedures.

1.4   Voltage definitions

The voltage definitions used throughout this document are those defined in BS 7671[1]. These are as follows:

— *extra-low voltage* (ELV): normally not exceeding 50 V AC or 120 V DC whether between conductors or to earth

— *low voltage* (LV): normally exceeding extra-low voltage but not exceeding 1000 V AC or 1500 V DC between conductors, or 600 V AC or 900 V DC between conductors and earth

— *high voltage* (HV): normally exceeding low voltage.

1.5   Other sources of information

It is hoped that this Guide will provide an invaluable reference source for those involved in the design, installation, commissioning, operation and maintenance of buildings when considering electrical services. However, it does not claim to be exhaustive. The various sections contained many references to other sources of information, particularly British Standards and associated standards and codes of practice, all of which should be carefully consulted in conjunction with this Guide, together with relevant trade and professional publications.

References

1   BS 7671: 2001: *Requirements for electrical installations. IEE Wiring Regulations. Sixteenth edition* (London: British Standards Institution) (2001)

2   *Inspection and testing* IEE Guidance Note 3 (London: Institution of Electrical Engineers) (2002)

2 Legislation and standards

2.1 Introduction to acts and legislation

The general installation and use of electricity in the UK is governed by various acts and regulations. These include the following:

— The Health and Safety at Work etc. Act 1974[1]

— The Electricity Act 1989[2]

— The Utilities Act 2000[3]

— The Electricity at Work Regulations 1989[4]

— The Building Standards (Scotland) Regulations 1990[5] (Technical Standards[6] Part N)

— The Electromagnetic Compatibility Regulations 1992[7]

— The Management of Health and Safety at Work Regulations 1999[8]

— The Building Regulations 2000[9] (Parts L and P)

— The Electricity (Standards of Performance) Regulations 2001[10]

— The Electricity Safety, Quality and Continuity Regulations 2002[11]

In addition to the above, many other regulations direct general safe working practices and have an impact on the installation and use of electricity. These include:

— Various construction regulations including the Construction (Design and Management) Regulations 1994[12] (under review)

— Health and Safety (Safety Signs and Signals) Regulations 1996[13]

— The Provision and Use of Work Equipment Regulations 1998[14].

— The Control of Substances Hazardous to Health Regulations 2002[15]

There are many more items of legislation regulating the way electricity is generated, distributed, operated, maintained and used. Much of the legislation is aimed at health and safety and those regulations made under the Health and Safety at Work etc. Act 1974 are enforced by the Health and Safety Executive (HSE).

Other regulations such as Occupiers Liability Acts (1957 and 1984)[16] make the occupier of premises responsible in common law for the safety of all occupants and visitors, even those intent on criminal activity.

The Electrical Equipment (Safety) Regulations 1994[17] are primarily aimed at consumer equipment being made under the Consumer Protection Act 1987[18]. However the boundary between consumer and professional electrical equipment is non-determinable and these Regulations have a wider impact than just consumer equipment.

Other legislation is enforced by the Department of Trade and Industry (DTI) such as the Utilities Act 2000 and the Electricity Act 1989 together with the Electricity Safety Quality and Continuity Regulations 2002.

2.2 British (BS), European (BS EN), International (ISO) and International Electrotechnical Committee (IEC) Standards

There are many standards relevant to electrical equipment and some are in the process of revision. Readers should refer to the British Standards Institution for the latest requirements (http://bsonline.techindex.co.uk).

2.2.1 BS 7671: 2001

BS 7671: 2001: *Requirements for electrical installations*[19] (previously referred to as the *IEE Wiring Regulations*) is, in itself, not a statutory requirement but is primarily a design standard.

Meeting this standard will ensure that the installation is installed, tested and maintained at a safe standard. It applies to systems known as low voltage, i.e. up to 1000 V AC. Any person involved with installation, testing and maintenance of low voltage systems must be thoroughly familiar and well practiced in its interpretation and application. Various bodies such as the Electricity Contractors Association (ECA) and the National Inspection Council for Electrical Installation Contracting (NICEIC) offer accreditations for suitably qualified persons.

2.3 Building Regulations 2000 and the Building Standards (Scotland) Regulations 1990

Under Part P of the Building Regulations 2000[9], compliance with BS 7671[19] has become compulsory for new domestic buildings in England and Wales. In Scotland this Standard has been legally binding for a number of years under The Building Standards (Scotland) Regulations 1990[5] (Technical Standards Part N[6]).

In order to approve the standard of completed work on all but minor alterations in England and Wales the person carrying out the work must have proof of competence in the form of an accreditation. Alternatively the local building inspector can be engaged to check compliance with BS 7671.

The Building Regulations 2000 (Part L in England and Wales) impacts of the use of energy in commercial and domestic buildings and requires effective energy management with the objective being energy conservation.

2.4   Regulations enforced by DTI

The Electricity Act 1988[2] and the Utilities Act 2000[3] determine the way in which the generation and distribution of electricity are structured and regulated in the UK. In England, Wales and Scotland the industry is regulated for the DTI by the Office of Gas and Electricity Markets (OFGEM). A similar body exists in Northern Ireland under a statutory order.

Amongst other requirements is the need for generation and distribution companies to operate under licence conditions. The Act deals with financial issues and metering, conditions for re-sale of electricity, complaints and the right of customers to certain information relating to their supply of electricity. They also oblige suppliers to utilise renewable energy.

The Electricity (Standards of Performance) Regulations 2001[10] have laid down performance standards, many of which are reflected in the standard licence conditions under which the companies operate. Speed of response to requests for supplies and alterations, the cost of alterations, the use of system charges are all determined by reference to these requirements. Some of these are soon to be revised.

As well as controlling standards of continuity and quality (e.g. voltage and frequency limits), the Electricity Safety, Quality and Continuity Regulations 2002[11] also control the safety of distribution equipment including cables and lines. It was originally intended that these Regulations would control distribution systems outdoors with BS 7671[19] being the design standard mainly for indoor supplies. However they strictly apply only to systems operated by 'distributors' (i.e. those who distribute electricity to another customer). In the modern world of multi-occupancy these Regulations are applying to a greater number of distribution systems than they would have done a few years ago.

The position and height of overhead cables are strictly regulated but the depth of cable significantly must be at a 'suitable depth'. Underground cables must be protected, clearly marked and recorded on maps.

These Regulations protect the public from dangers within substations. Because of public perception relating danger to voltage levels, the Regulations have abandoned 'Danger' signs relating to voltage and now enforce the use of a yellow sign, complying with the Health and Safety (Safety Signs and Signals) Regulations 1996[13], stating the clear message 'Danger of Death' and accompanied by a pictogram showing the electrical flash and a falling person.

2.5   The Health and Safety at Work etc. Act 1974

The Health and Safety at Work etc. Act 1974[1] places obligations on management and employees to operate in a safe manner and to co-operate to ensure safety standards. It is policed by the Health and Safety Executive. The Act controls all risks and ensures good health at work and standards of welfare. It is an enabling Act and has enabled the formulation of regulations such as the Electricity at Work Regulations 1989.

Breach of the Act is a criminal offence and fines of up to £20 000 can be imposed by magistrates plus up to six months' imprisonment. A Crown Court can impose unlimited fines and up to 2 years' imprisonment.

Section 2 of the Act places stringent duties on employers and Section 3 places similar duties on contractors and other not in employment. Section 4 controls the safety of premises. Sections 7 and 8 relate to employees.

2.5.1   The Electricity at Work Regulations 1989

The Electricity at Work Regulations[4] place responsibilities on 'duty holders' to ensure that safety is maintained at all times. Duty holders are defined as anyone who has a duty under the Regulations. Duty holders can be directors, company secretaries, management and employees with responsibilities for the interpretation and maintenance of safety standards to do with the installation, appliances, work procedures and other elements.

Regulation 4 governs safe working practices and regulation 5 governs the system hardware and all equipment in use or available for use. Of particular interest are:

—   Regulation 13 enforces safe working practices to be implemented before work can commence on any equipment made dead. It also enforces safe working procedures in making the equipment dead and the HSE's *Memorandum of Guidance on the Electricity at Work Regulations 1989*[22] stipulates that any equipment over 3000 volts, or complex equipment at any voltage, must be covered by house rules or other written procedures. This rule is seen to enforce the need for mandatory company safety rules and procedures and specific step-by-step guides such as 'switching' or 'safety' programmes.

—   Regulation 14 is a three-part strict and absolute requirement before any work can be carried out live.

—   Regulation 16 covers the need for competency and effective supervision.

2.5.2 Management of Health and Safety at Work Regulations 1999 and various construction Regulations

Also under the umbrella of the Health and Safety at Work etc. Act 1974[1] are the Management of Health and Safety at Work Regulations 1999[8] (MHSWR) and various construction regulations including the Construction (Design and Management) Regulations 1994[12] (CDM) (soon to be completely revised).

The MHSWR require risk assessments to be made for every aspect of the work and these must be written down if five or more people are employed. They also require safe working practices to eliminate the dangers identified. A new requirement under the 1999 version of the MHSWR is regulation 8(b) which prohibits persons from rooms that contain dangers (e.g. plant rooms and substations) unless they have adequate knowledge and training to avoid danger. This means that such rooms must remain locked with access restricted to those deemed to be competent. The Regulations were introduced in 1992 and were heavily revised in 1999 to enable the UK to comply with various European Directives affecting safety, health and welfare.

The CDM Regulations are particularly stringent governing the management of construction works, including works on electrical plant and mains.

2.5.3 Other miscellaneous regulations

Other sets of regulations include the Control of Substances Hazardous to Health Regulations 2002[15], impacting on the use of substances such as mineral switch oil, PCBs, SF6 gas and other potentially hazardous substances. The Control of Asbestos at Work Regulations 2002[20] and The Control of Lead at Work Regulations 2002[21] and their associated licensing conditions may also be appropriate. The Provision and Use of Work Equipment Regulations 1998[14] ensure that such equipment is suitable for the task and well maintained. The Confined Spaces Regulations 1997[23] may also apply.

2.6 The Electromagnetic Compatibility Regulations 1992, European Harmonisation Directives and other European Directives

Directives are issued by the EC legislature and the UK is obliged to comply with their requirements. Central to the Directives are:

— to ensure a common standard of safety at work across the European Community

— by harmonisation of delivery (e.g. voltage standards) and equipment, to enable equipment to be used across the European Community.

Most of these framework and individual directives are made to establish a common European market where goods and services can be prepared in one nation and sold or deployed in another. Thereby equipment will work anywhere in Europe.

Certain UK regulations have reflected this common ground in UK law, for instance the Management of Health and Safety at Work Regulations 1999[8] (as did their 1992 predecessor) enforce safety standards where the UK was deemed to fall short.

The Low Voltage Directive[24] enforces a common (low) voltage across Europe. In the UK this is currently stated as 230 V/400 V +10%/−6%, enforced by the Electricity Safety, Quality and Continuity Regulations[11].

The Electromagnetic Compatibility Regulations 1992[7] seek to impose standards by which one piece of equipment will not interfere with another. Electromagnetic compatibility is considered in detail in section 11 of this Guide.

Compliance with the Harmonisation Directives and regulations enables the attachment of the 'CE' label to show that the equipment is suitable for use across Europe.

References

1   The Health and Safety at Work etc. Act 1974 (London: Her Majesty's Stationery Office) (1974)

2   The Electricity Act 1989 (London: Her Majesty's Stationery Office) (1989)

3   The Electricity at Work Regulations 1989 Statutory Instrument 1989 No. 635 (London: Her Majesty's Stationery Office) (1989)

4   The Building Standards (Scotland) Regulations 1990 Statutory Instruments 1990 No. 2179 (S. 187) (London: Her Majesty's Stationery Office) (1990)

5   Technical standards for compliance with the Building Standards (Scotland) Regulations 1990 (as amended) (London: The Stationery Office) (2001)

6   The Electromagnetic Compatibility Regulations 1992 Statutory Instruments 1992 No. 2372 (London Her Majesty's Stationery Office) (1992)

7   The Management of Health and Safety at Work Regulations 1999 Statutory Instruments 1999 No. 3242 (London: The Stationery Office) (1999)

8   The Building Regulations 2000 Statutory Instruments 2000 No. 2531 (London: The Stationery Office) (2000)

9   The Electricity (Standards of Performance) Regulations 2001 Statutory Instruments 2001 No. 3265 (London: The Stationery Office) (2002)

10   The Utilities Act 2000 (London: The Stationery Office) (2000)

11   The Electricity Safety, Quality and Continuity Regulations 2002 Statutory Instruments 2002 No. 2665 (London: The Stationery Office) (2002)

12   The Control of Substances Hazardous to Health Regulations 2002 Statutory Instruments 2002 No. 2677 (London: The Stationery Office) (2002)

13   The Construction (Design and Management) Regulations 1994 Statutory Instruments 1994 No. 3140 (London: Her Majesty's Stationery Office) (1995)

14   The Health and Safety (Safety Signs and Signals) Regulations 1996 Statutory Instruments 1996 No. 341 (London: Her Majesty's Stationery Office) (1996)

15   The Provision and Use of Work Equipment Regulations 1998 Statutory Instruments 1998 No. 2306 (London: The Stationery Office) (1998)

16   Occupiers Liability Act 1957 and Occupiers Liability Act 1984 (London: Her Majesty's Stationery Office) (1957/1984)

17   The Electrical Equipment (Safety) Regulations 1994 Statutory Instruments 1994 No. 3260 (London: Her Majesty's Stationery Office) (1994)

18   The Consumer Protection Act 1987 (London: Her Majesty's Stationery Office) (1987)

19   BS 7671: 2001: *Requirements for electrical installations. IEE Wiring Regulations. Sixteenth edition* (London: British Standards Institution) (2001)

20   Control of Asbestos at Work Regulations 2002 Statutory Instruments 2002 No. 2675 (London: The Stationery Office) (2002)

21   The Control of Lead at Work Regulations 2002 Statutory Instruments 2002 No. 2676 (London: The Stationery Office) (2002)

22   *Memorandum of Guidance on the Electricity at Work Regulations 1989* Health and Safety Executive HS(R)25 (London: Her Majesty's Stationery Office) (1998)

23   The Confined Spaces Regulations 1997 Statutory Instruments 1997 No. 1713 (London: The Stationery Office) (1997)

24   Directive 73/23/EEC of the European Council of 19 February 1973 on the harmonisation of the laws of Member States relating to Electrical Equipment designed for use within certain voltage limits *Official J. of the European Communities* 26.03.1973 L77/29 (Brussels: Commission of the European Communities) (March 1973)

3   Load assessment

3.1   Introduction

Modern buildings contain an increasing amount of electrical apparatus. This technological explosion is increasing the demand for electrical power to support the equipment which has become an integral part of the way people conduct business. This section outlines the basic items for consideration when evaluating the electrical load for a commercial building. Operating load requirements must be estimated realistically if a properly engineered electrical distribution system is to be achieved.

Detailed information about the equipment to be installed in a building, whatever its intended use, is unlikely to be available at the early planning stages.

In order for planning to proceed, utility companies to be contacted and decisions to be taken so as to enable major elements of plant and service equipment to be specified, the design engineer must be able to confidently produce load estimates. These estimates should be based upon experience and a broad database of actual operating loads for similar projects.

3.2   Definitions

3.2.1   Active power (useful power)

Active power, also known as real power, is the time average of instantaneous power when the average is taken over a complete cycle of an AC waveform. This is expressed in watts (W).

The active power when the voltage and current waveforms are sinusoidal can be calculated by the following.

For single phase:

$$P = V I \cos \phi \tag{3.1}$$

where P is the active power (W), V is the voltage (V), I is the current (A), ϕ is the angular displacement between the voltage and current waveforms and is known as the power factor angle; $\cos \phi$ is known as the power factor.

For balanced three-phase:

$$P = \sqrt{3}\, V_{(ph-ph)} I \cos \phi \tag{3.2}$$

where $V_{(ph-ph)}$ is the phase-to-phase voltage (V).

3.2.2   Apparent power

Apparent power is the product of RMS voltage and current, expressed in volt-amperes, and is calculated as follows.

For single-phase:

$$AP = V I \tag{3.3}$$

For three-phase:

$$AP = V_{(ph-ph)} I \sqrt{3} \tag{3.4}$$

where AP is the apparent power (volt-ampere).

This is often the basis for rating the electrical requirement. This traditional definition is based upon sinusoidal waveforms. For non-sinusoidal waveforms other characteristics of the voltage and current wave shape may need to be considered when rating equipment. For example, the peak voltage levels or current levels may be considerably higher than those for a sinusoid of the same RMS value. These voltages and currents can contribute to excessive insulation stress, thermal stress etc.

3.2.3   Connected load

The sum of all the loads connected to the electrical system; usually expressed in watts.

3.2.4   Demand factor

This is the factor applied to the connected load of an installation to assess the maximum demand value. This will take account of the fact that not all installed equipment is used at full load or at all times. The demand factor is always less than or equal to unity, and is given by:

$$\text{Demand factor} = \frac{\text{Maximum demand}}{\text{Connected load}} \tag{3.5}$$

3.2.5   Load factor

Load factor is the ratio of the average demand to the maximum demand over a defined interval. In most commercial office buildings, for example between the hours of 09:00 to 18:00 the load factor is usually close to unity. A typical load profile for a commercial office building is graphically represented in Figure 3.1.

The maximum demand, or demand load, is the actual operating load (in kW) of a system. The demand load may be measured using a wattmeter or calculated by multi-

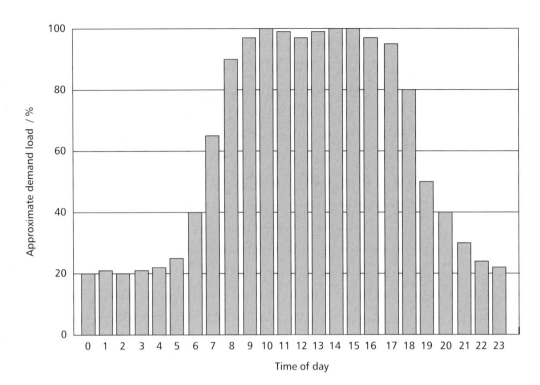

Figure 3.1 Typical load profile for an office building

plying the connected load by the demand factor. For the purposes of revenue metering for electrical energy suppliers, maximum demand is averaged over a small increment of time, usually over 30 minutes, or sometimes over 15 minutes.

3.2.6 Power factor

Power factor is defined as the ratio of the apparent power in a circuit (V·A) to the useful power (W) if the voltage and current are sinusoidal; i.e. power factor = kW/kV·A.

3.3 Load assessment

The electrical load within most commercial buildings can be arranged into the following broad categories:

— lighting

— small power and special user equipment

— heating, ventilating and air conditioning (HVAC) equipment

— lifts and escalators.

The electrical demand is directly connected to the heat output and this is discussed in section 6 of CIBSE Guide A: *Environmental design*[1].

3.3.1 Lighting

The design of lighting systems is provided in the CIBSE *Code for lighting*[2] and CIBSE/SLL Lighting Guides.

Case studies carried out on a number of offices built or refurbished since 1983 found that the lighting loads were between 8 and 18 W·m^{-2} for maintained illuminance levels of 350–500 lux.

In some areas of a building special-purpose lighting is likely to result in higher electrical loads greater than this. Special-purpose lighting includes theatrical lighting, display lighting in retail establishments, lighting for medical treatment areas and operating theatres etc. The impact of these types of lighting should be evaluated on a case by case basis. However, areas requiring speciality lighting, when part of a large office building, may represent a small fraction of the total floor area and therefore may have a small or negligible effect on the overall electrical capacity required from the lighting system.

For broad planning purposes, taken over the whole of a large office building, a unit load of 12–20 W·m^{-2} is reasonable.

3.3.2 Small power and special user equipment

Small power and user equipment generally consists of items which are plugged into socket outlets or permanently connected. Most of this type of equipment is brought into the building by the occupants and not selected by the building services designer.

The small power requirements vary widely throughout a building, from some areas having virtually no small power loads to other areas, such as financial trading rooms or computer rooms, which have a relatively high unit loading.

If special areas exist with known loads, a higher average load should be used for these areas. For example, equipment loads in excess of 250–500 W·m^{-2} and 200–300 W·m^{-2} may be required for a large computer room or a financial dealing area, respectively. It is of extreme importance that the engineer obtains details of all the connected equipment and includes an appropriate allowance for future expansion or load increase.

3.3.3 Heating, ventilating and air conditioning equipment

In modern ventilated and air conditioned buildings the load required for HVAC apparatus can represent 40–50% of the total building load. Such loads are affected by the nature of the building fabric, fresh air requirements and internal heat gains from lights, people and equipment, and should be determined by the project mechanical engineer. Typically, the electrical load resulting from a mechanically ventilated and cooled building could be in the order of 40–50 W·m^{-2}, but requirements must be assessed for each project.

In electrically heated buildings, the required load must be similarly calculated by the project mechanical engineer. Since the heating and cooling loads are not usually simultaneous, only the larger of the two should be used for system calculation.

3.3.4 Lifts

The evaluation of lift requirements must be undertaken by a lift specialist who, on the basis of the building population, should determine the number, speed and capacity of lifts. Further guidance on the planning and design of lifts is provided in CIBSE Guide D: *Transportation systems in buildings*[3].

A large, tall building may have several hundred kilowatts of lift equipment installed. Of particular interest to the electrical engineer, however, is the fact that peak lift loads can be considered short time loads and their impact on overall building demand discounted (but not ignored).

The starting current of a large lift motor must be evaluated for potential effects on system voltage. Additionally, the nature of solid state speed controllers commonly used today may create disturbances on the distribution system, particularly when the lift load is a large percentage of system capacity as in a small building or when operating on a standby generator.

3.4 Load calculations

An evaluation of the main electrical services requirements in a building should begin with an estimate of the likely load requirements. This is usually based upon a unit loading on a square metre basis. The unit loading used should be based upon a reasonable figure, statistically determined from an analysis of similar buildings.

When applying unit loads on a W·m^{-2} basis for initial design calculations, it is suggested that the gross area of the building be utilised, subtracting the known areas of lifts, shaftways, etc. Reducing areas to take account of the thicknesses of exterior walls or columns is an unnecessary level of precision for estimating loads. The difference between lettable areas and building gross areas has little impact on equipment selection, unless unreasonably large unit loads are used for the calculations.

Table 3.1 summarises unit load estimates for lighting and small power for various types of building. An estimate for

Table 3.1 Minimum design load capacities for lighting and small power equipment for various types of building

Building type	Minimum load capacity / (W/m^2)
Office	60
School	30
Residential building	30
Hospital	25
Hotel	25
Church	15

Table 3.2 Measured total maximum demand loads for representative sample of large office buildings

Building	Gross area / m	Maximum demand / MW	Maximum demand / W·m^{-2}
A	35 000	2.32	67
B	28 000	2.34	84
C	40 000	3.28	82
D	46 000	2.42	53
E	50 000	2.1	42

mechanical apparatus such as lifts, air conditioning equipment, pumps etc., should be added to each of these figures. These loadings are recommended minimum requirements.

Table 3.2 presents a summary of measured maximum operating demand loads for modern 'high-tech' commercial office buildings. Each building is fully air conditioned and contains dealing facilities of various sizes, some accommodating as many as 250 traders. It is significant to note that the highest recorded demand is considerably below some recent user expectations of 150–200 W·m^{-2}.

3.5 Harmonics

All electrical installations must comply with the requirements of Electricity Association Engineering Recommendation G5/4[4], which imposes limits on the extent of harmonics that can be generated in an electrical system.

Harmonics are typically generated by inductive luminaires, computers, rotating machines, electronic speed controllers, arc furnaces and uninterruptible power supplies (see section 9).

Harmonic currents may cause distortion of the voltage waveform (known as notching) and high neutral currents. The notching effect on the voltage waveform can be particularly troublesome on small distribution systems (i.e. high-impedance systems such as standby generators) as a result of instantaneous voltage drop occurring with each pulse of input current as solid state switching devices fire. This can cause equipment malfunction due to line voltage conditions being outside accepted tolerances. It is essential for the equipment designer to have an understanding of load characteristics so that provisions may be designed into the system to accommodate the equipment.

Where possible, the electrical engineer should persuade those responsible for the purchase of equipment to specify maximum permissible levels of input current harmonics. In addition, the manufacturer of the equipment should be required to provide suitable filters, input isolation transformers etc., as needed to maintain the permitted harmonic levels.

High neutral currents may be present in systems containing harmonic current due to the fact that third order harmonic current components in a three-phase, four-wire system (TPN) may become arithmetically additive in the neutral. For this reason, multicore cables with reduced neutrals should not be specified for installations where significant harmonic currents are likely to occur.

3.6   Codes and standards

There are no regulations in the UK for minimum design capacities for various buildings; this is left to the judgement of the electrical designer. This is not the case in all inspection jurisdictions. Where the engineer is designing an electrical installation for use in another country, codes or standards may require a minimum stated capacity in order for a building to be considered adequate for its intended purpose. Therefore, the engineer should determine the following early on in assessing load requirements:

—   What codes and standards apply?

—   What regulations, if any, dictate minimum provisions for electrical capacity?

—   What local practices and conventions apply, regarding earthing, cable calculations, over-current and short circuit protection etc?

3.7   Diversity and demand

3.7.1   General

The actual operating load rarely, if ever, equals the sum of all the loads installed. The maximum operating load can be related to connected loads or to the sum of demand loads by the demand factor or the diversity factor. To avoid confusion designers must always keep in mind actual operating conditions in an installation. An electrical distribution system can be broken down into groups of smaller systems or branches, successively connected together forming the whole network. Each branch of the network can contain smaller branches and individual items of equipment. The operating load existing at any location in a system at a given point in time is the sum of the loads downstream existing at that time. Diversity occurs in an operating system because not all loads connected are operating simultaneously or are not operating simultaneously at their maximum rating. Examples of this occur in a building as follows:

—   Lighting is often used in response to the amount of daylight or to occupancy, particularly if automatic controls have been installed. It is unusual for all the lighting to be operated at the same time.

Store rooms and other ancillary areas are often left unlit for long periods.

—   Equipment connected to socket outlets or fused connection units are often used intermittently. This would apply to such equipment as photocopiers or hand driers.

—   Motors do not usually operate at their nameplate rating, i.e. the mechanical load on the rotating shaft is less than the rated load. This could be due to operating variations in load, as in an air conditioning system with variable air volume controls or due to conservative equipment selection by the designer.

—   Some loads are cyclic in nature, as with sump pumps, sewage ejector pumps, air compressors, lifts, etc. This means that probably not all loads will be operating simultaneously.

—   Some loads rarely operate, except in unusual circumstances or for testing, such as in the cases of sprinkler booster pumps or hose reel pumps.

—   Once they have reached a set operating temperature many items of electrical cooking equipment draw full load only for short periods to retain this level.

—   Other equipment often operates at less than manufacturers' nameplate ratings.

The diversity factor recognises that the load does not equal the sum of its parts due to time interdependence (i.e. diversity). It is defined as the ratio of the sum of the individual demands in a system to the maximum demand for the system. It is always equal to or greater than unity. A diversity factor of unity indicates that all loads are coincident. An example would be a conveyor belt made up of six sections, each driven by a 2 kW motor. As material is transported along this belt, it is initially carried by the first section, then by each section in succession until the final section is reached. In this simple example only one section of conveyor is carrying material at any moment. Therefore five motors are handling no-load mechanical losses (say 0.1 kW), keeping the belt moving, whilst one motor is handling the load (say 1 kW). The demand presented by each motor when it is carrying the material is 1 kW and the sum of the demand loads is 6 kW, but the maximum load presented by the system at any time is only 1.5 kW.

The diversity factor for this system is given by:

$$\text{Diversity factor} = \frac{\sum (\text{Demands})}{\text{Maximum demand}} \quad (3.6)$$

Therefore, for the above example:

Diversity factor = 6 / 1.5 = 4

From equation 3.5, the demand factor for the system is:

Demand factor = 1.5 / (2 × 6) = 0.125

If, however, the carrying rate of the conveyor is increased such that each section is carrying a load, the maximum load would increase to 6 kW. Therefore the diversity factor would decrease to unity and the demand factor would increase to 0.5.

Many building services engineers find it convenient to work with the demand factor and will often express the effect of diversity as a percentage of connected load.

3.7.2   Lighting

In commercial or industrial buildings, lighting demand is often assumed to equal 100% of the connected load. However, some luminaires will not be operating. In installations provided with local switching, daylight linking (or occupancy-sensing) lights in unoccupied areas may not be on. The extent of this diversity may be very small but nonetheless it exists.

3.7.3   Small power

Recently, there has been a trend to anticipate the use of large quantities of electronic equipment in general office spaces. As a result design levels for small power loads have been rising. Experience has shown, however, that demand loads have not grown as rapidly as connected loads. This may be due to several factors, including the following:

— overestimating loads: in some cases the quantity of equipment anticipated is never actually installed

— under-estimating or neglecting the effect of diversity and demand factors.

The design engineer must have an understanding of realistic values for operating demand as a percentage of connected loads. This understanding can be assisted by good record keeping and gathering operational data from completed installations.

Small power loads can significantly affect the capacity of air conditioning plant. A reasonably accurate assessment of operating demand loads will enable an HVAC design engineer properly to select equipment for a project. Oversized equipment can be extremely difficult to control when operating at very small percentages of rated capacity. In some cases, as in variable air volume (VAV) systems, occupied spaces may be severely over-cooled even at minimum control settings.

3.7.4   Power factor

In an operating system various loads comprising the system may have differing power factors. This must be considered when totalling loads at any point in the system, and loads should be added algebraically. It is often convenient for initial analysis to work in real loads (in kW) only.

3.7.5   Electric heating

When a building is electrically heated, the heating load must be considered independently from any unit load design guidelines and generally added to the maximum demand calculations. In an air conditioned building, however, it is likely that maximum heating load will not occur simultaneously with cooling loads. Therefore, this diversity must be considered in determining maximum demand.

3.7.6   Voltage variation

In general, all load evaluation is based upon nominal system voltage. The designer should always be aware of the effects of voltage drop or variations in supply. Voltage drops within the range recommended by various codes and standards or supply company regulations will not usually present any problems.

References

1   *Environmental design* CIBSE Guide A (London: Chartered Institution of Building Services Engineers) (1999)

2   *Code for lighting* (London: Chartered Institution of Building Services Engineers) (2002)

3   *Transportation systems in buildings* CIBSE Guide D (London: Chartered Institution of Building Services Engineers) (2000)

4   *Planning levels for harmonic voltage distortion and the connection of non-linear equipment to transmission systems and distribution networks in the United Kingdom* Engineering Recommendation G5/4 (London: Electricity Association) (2001)

4 Off-site supplies and on-site power generation

4.1 Off-site supplies

4.1.1 Introduction

Prior to the introduction of The Electricity Act 1989[1], electricity was generated and distributed in England, Wales and Scotland by the Central Electricity Generating Board (CEGB). The CEGB sold electricity in bulk to 12 area distribution boards, each of which served a closed supply area or franchise. A co-ordinating body, the Electricity Council, dealt with overall policy matters.

The 1989 Electricity Act[1] laid the legislative foundations for the restructuring and privatisation of the electrical supply industry whereby the CEGB was split into three generating companies and a transmission company. The power stations were divided between two fossil-fired generating companies, National Power and PowerGen and a nuclear generating company, Nuclear Electric.

National Power and PowerGen became private companies in 1992 while Nuclear Electric was privatised as part of British Energy in 1996.

The ownership and operation of the transmission system were transferred in 1990 to the newly created National Grid Company as the transmission system operator. National Grid was responsible for the transmission of electricity to the Regional Electricity Companies (RECs). National Grid was jointly owned by the Regional Electricity Companies but was floated on the stock market as an independent company in 1995.

Twelve Regional Electricity Companies were created as the successors to the previous area boards and were privatised in December 1990. The RECs are responsible for the distribution of electricity over their network and the supply of electricity to final customers.

In Scotland, vertical integration was maintained in the new structure with the creation of Scottish Power and Scottish Hydro-Electric. As in England and Wales, nuclear generation was assigned to a separate company, Scottish Nuclear, which became part of British Energy in 1996.

In Northern Ireland, Northern Ireland Electricity became responsible for transmission, distribution and supply.

The 1989 Electricity Act[1] also created a system of independent regulation. In 1999, the Regulatory Offices for electricity and gas in England, Wales and Scotland were merged to form the Office of Gas and Electricity Markets (OFGEM). Northern Ireland has it own Regulatory body, the Office for the Regulation of Electricity and Gas (OFREG).

4.1.2 Electricity transmission

The UK transmission system, the National Grid, was established in the 1930s and was designed to operate at 132 kV. In the early 1950s, the operating voltage was raised to 275 kV and in 1960 the CEGB decided to introduce new transmission lines operating at 400 kV.

Figure 4.1 shows generating and transmission arrangements within the National Grid.

The power dissipated in a resistance (such as a length of transmission cable) is proportional to the square of the current passing through the resistance (i.e. $P = I^2 R$). Therefore, the lower the current passing through the cable, the lower the losses generated in that cable which are dissipated as heat.

Since power is the product of voltage and current, these losses can be reduced by the adoption of higher trans-

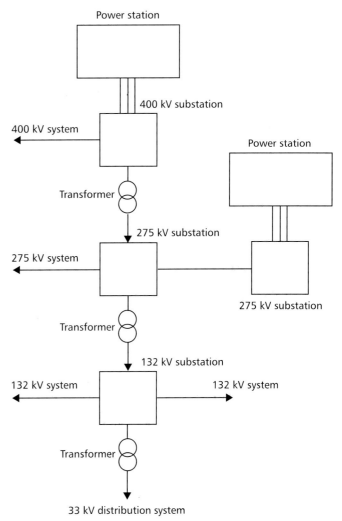

Figure 4.1 Generating and transmission arrangements within the National Grid

mission voltages. However, the higher transmission voltages increase the risk of flashover between adjacent cables or between the cables and the supporting structures. In the design of transmission systems, a balance must be struck between acceptable transmission losses and the high cost of HV transmission equipment.

4.1.3 Quality of electrical supply

The load on the National Grid varies considerably during the day and from season to season. The generating and distribution companies have a statutory obligation to satisfy the load and, at the same time, maintain the quality of the power supply within 1% of the declared supply frequency and within +10% and –6% of the declared supply voltage. The quality of the electricity supply is defined in the Electricity Safety, Quality and Continuity Regulations 2002[2].

4.1.4 Wholesale electricity market

All generating companies are required to sell the electricity they produce into an open commodity market, known as the Pool. Each generating company has to declare in advance its availability to the market, together with a price at which it is prepared to generate. From this information, the National Grid Company establishes a schedule of generating plant in ascending order of price. The Pool has been overlaid with contracts, short and long term, to reduce volatility in the market and to make capacity and energy prices more predictable.

The RECs buy electricity in bulk from the Pool and, in turn, sell the electricity to consumers at a voltage appropriate to the load of the consumers installation. For a large industrial site, the electricity may be supplied at a voltage of 33 kV.

For commercial or industrial premises where the load exceeds approximately 1 MW, the supply voltage would be 11 kV. The smaller commercial or industrial premises, electricity would be provided as a 3-phase low voltage (400 V) supply. Domestic premises would be supplied with a single phase (230 V) supply.

4.1.5 Tariffs

The RECs offer a range of tariffs to suit a variety of consumer types. For domestic consumers, the choice is generally between a standard rate with all units charged at the same value and an off-peak rate with low cost units during night time hours.

For non-domestic users, the most widely used tariff is the maximum demand tariff which is a complex pricing system consisting of several different components:

— a standing monthly charge

— an availability charge, based on the maximum anticipated electrical load of the installation

— a demand charge for each kilowatt of the maximum demand in each month; this is generally small for the summer months when demand on the supply is light but increases considerably

during the winter months when electric heating demands are high

— a charge for the number of units (kilowatt-hours) used; this charge may vary between units consumed during the day and those used at night.

In general, a supply provided by a REC at high voltage (HV) will attract lower tariff charges than those provided at low voltage (LV). This is a reflection of the reduced amount of plant that has to be provided and maintained by the RECs for high voltage consumers.

4.1.6 Incoming electrical supplies

Regional electricity companies have a variety of guides that are freely available to designers to assist in the procurement of potential supplies for projects.

As a prerequisite to providing a supply, the REC would require basic information about the project such as the location, size and type of development, the type of load and the maximum power required. In addition, the REC will wish to know the method of heating the building.

When the incoming supply is small, the building can be connected to the existing LV network without the need for reinforcement of the supply network. For large building loads, it will be necessary to establish a new substation to serve the project and alteration of the HV network may be required. The need for a new substation will depend on a variety of factors including the location of the site (i.e. rural or urban), the nature of the project and the existing load on the REC network.

It is important that the maximum power requirement for the project is assessed as accurately as possible. Too low a figure could result in the rating of the supply equipment being insufficient to meet the actual power requirements. However, if the assessed value of maximum power is set too high, this will result in unnecessarily high availability charges being levied by the REC.

The substation may be owned and operated by the REC, providing the user with a LV supply, or the supply may be provided by the REC at high voltage (11 kV) with the substation owned and operated by the building owner.

4.1.7 Substation planning

There are usually three main electrical components contained in a substation. These are HV switchgear, one or more transformers and LV switchgear. These may be contained in a single area or enclosure or may be segregated into separate rooms, depending on the circumstances of the building. In general, REC substations are contained within a single room.

As an approximate guide, the internal space requirements for substations are as follows:

— substation containing a single transformer: 4500 mm × 3500 mm × 2800 mm high

— substation containing two transformers: 6700 mm × 4500 mm × 2800 mm high

— high-voltage switch room: 4000 mm × 2200 mm × 2800 mm high.

The dimensions given are approximate and are the minimum required. Larger supplies will inevitably necessitate greater space requirements to house the equipment.

In addition, space will be required to install the LV switchgear and this should be positioned in a switch room immediately adjacent to the substation or transformer chamber.

Layouts for typical substations are shown in section 7.

4.1.8 Wayleaves and easements

Where a REC installs cables or equipment as part of a project, there will be a requirement to negotiate a wayleave or easement to ensure that the REC has a legal right to gain access to the equipment in the future. It is normally a condition of the agreement for a new supply that wayleaves/easements are granted under the terms as agreed, and that all documentation is signed before electricity is provided through the supply connection. The term of the lease is not normally less than 99 years.

It is a standard requirement of all RECs that each substation on the REC network is provided with a means for 24-hour personnel access for switching off that part of the network and with suitable vehicle/craneage access to the installed plant during working hours.

4.1.9 Substation construction

Many issues require consideration in the design of substations, whether or not they are owned and operated by the REC or by the building owner.

The substation needs to be securely constructed where it is enclosed and, when necessary, fire separation must be provided between the substation and the adjacent accommodation to meet the requirements of the Fire Officer and other regulatory bodies.

Equipment within the substation can produce a considerable amount of heat, necessitating adequate ventilation to maintain the environment within the substation below the condition at which equipment and cables need to be derated. Ideally, the ventilation should be configured to be natural, thus avoiding maintenance of mechanical ventilation systems, and to achieve this, the substation should be situated close to a source of fresh air.

Where a substation is located outdoors, adequate protection in the form of vandal-proof fencing with lockable gates should be provided to avoid unauthorised access. In this situation, the equipment specified must be fully weatherproof. New requirements for external substations have been introduced in The Electricity Safety, Quality and Continuity Regulations 2002[2].

The substation needs to be designed to ensure the easy installation and possible future replacement of cables. Where the cables enter the switchgear or transformer from below, trenches and/or ducts of appropriate dimension and depth should be installed. Consideration needs to be given to the bending radius of each cable (see section 8.6.1) that is to be installed to ensure that, at the depth proposed, it

can be terminated on the glanding plates of the switchgear or transformers.

The electrical installation at each substation requires earthing to meet the requirements of the REC and BS 7430[3]. This may be carried out using rods driven into the ground or by a coarse mesh made up from buried copper tape. The subject of earthing is discussed in detail in section 10.

4.2 On-site power generation

4.2.1 Introduction

There are clear a number of fundamental reasons for considering the provision of private or standby generation of electricity, independent from the incoming REC supply. Three of the most common of these are:

— *standby generation*: if the particular application demands an alternative source of supply to protect against failure or disruption of the public supply

— *private generation*: if a total independence from the public supply is required; either for security or economic reasons

— *combined heat and power* (CHP): where electricity is produced as part of a CHP scheme.

4.2.1.1 Standby generation

This is commonly used where it is necessary to maintain a secure supply for special equipment such as computers, communications or similar high risk systems, or for safety purposes to maintain supplies for firefighting equipment, other life safety systems, lifts in high-rise buildings, water and sewage pumps and services.

When selecting loads to be supported by standby generation it is necessary to study each item of plant and decide which are absolutely essential.

Should one set of a multi-set installation fail there must be a control programme immediately available to reduce the load to the remaining sets.

4.2.1.2 Private generation

The RECs have adopted a flexible approach, such that private generation has become attractive, both for peak lopping and base load purposes. Should there be a requirement for heating, then an integrated CHP system (utilising the waste heat from the engines or turbines) may be adopted which would increase the thermal efficiency of the installation from approximately 40% to 80% for a reciprocating engine and 28% to 80% for a gas turbine, thereby improving the economics of private generation.

For base load with reciprocating engines, it would be ideal to install two extra sets, one for on-line/off-line reserve, and one out of service for maintenance.

In order to reduce the number of sets, generator failure could be covered by connection to the REC supply. An important factor is the cost of such a connection and has to be included in the evaluation. Financial benefits may be gained by exporting to the grid when possible, making it necessary to run in parallel with the supply. Such arrangements will require appropriate controls and agreements with the REC and all customers wishing to install private generation must comply with the Electricity Association's Engineering Recommendation G59/1[4].

The Electricity Association has also produced Engineering Technical Report ETR113: *Notes of guidance for the protection of private generating sets up to 5 MW for operation in parallel with PES distribution networks*[5].

4.2.1.3   Combined heat and power

Combined heat and power (CHP) schemes offer a significantly greater efficiency than conventional power stations, typically converting 25–35% of the fuel energy to electricity and 50–60% as heat. CHP units can run on a variety of fuels including natural gas, bio-gas and diesel oil.

Electrically, CHP units should be treated as private generators as described in 4.2.1.2.

Further guidance on CHP systems is given in CIBSE Guide F: *Energy efficiency in buildings*[6].

4.2.1.4   Temporary/mobile sets

In cases where temporary power is needed suitable sets can either be purchased or hired. These include:

—   hand-portable units

—   hand trolley units

—   trailer-mounted units

—   skid-base units

—   low-load transportable units.

Starting can be by hand start, electric hand start or automatic start.

The arrangement for connecting these types of set to the site load vary. The smaller hand-portable or hand trolley sets will normally have their own panel with simple plug-in connection as part of the unit. The larger mobile sets may have their own control and protective equipment as an integral part of the unit. However, in some instances this may need to be provided separately. Care should be taken that proper protection and safety measures are provided in all instances.

4.2.2   Prime mover

The main power source is termed the prime mover, and for standby or private generation the most commonly used at present are:

—   internal combustion reciprocating engine, either diesel, gas or dual fuel

—   gas turbine

—   steam turbine

—   wind generator.

In addition to the above, water wheels and water turbines could provide the source of the motive power.

4.2.2.1   Prime mover selection

The selection of prime mover is governed by:

—   physical size

—   weight

—   location

—   choice of fuel

—   performance

—   capital and running costs.

4.2.2.2   Reciprocating engines

In order to obtain maximum power output from reciprocating engines it is necessary to provide turbocharged engines, either with or without air coolers (intercoolers). The latter reduces the size of the engine and hence the space required.

Sets operating at 1500 rev/min and above are normally selected for standby duty. These higher speeds may be utilised for base load applications where restricted space and power-to-weight ratios are important. To meet these requirements turbines should also be considered.

Reciprocating engines of up to 10 MW should start up and accept load in approximately 10 to 20–30 seconds. Acceptable steps in load should be discussed with the manufacturer and are dependent on the type of engine.

4.2.2.3   Gas turbines

It is uncommon for gas turbines to be considered for loads below 500 kW. Starting times are normally around 45–60 seconds, but rated full load can normally be accepted in one step when the turbine is up to running speed.

Single shaft turbines are generally considered the best option for standby purposes and in situations subject to large load variations.

4.2.2.4   Wind generators

In general the design and application of wind generators is complex and involves aerodynamics, mechanics and geography in addition to electrical engineering. Wind energy is freely available, avoids fuel transportation problems and is virtually inexhaustible, but cannot be completely relied upon. There is a wide range in application size and can it therefore be used for peak lopping and large distribution systems.

When site conditions are being considered, it should be noted that the power output is proportional to the cube of the windspeed. Therefore all aspects relating to its location are vital, the height above the ground, the proximity to any obstruction, the noise generated and the size of unit. If the unit is to be mounted on top of a building, it will be necessary to investigate the effect the aerodynamic forces will have on the structure.

Information relating to the wind pattern in the particular area covering wind speed and time can be obtained from the British Wind Energy Association*.

4.2.2.5 Prime mover rating

The declared power output of a prime mover is the power available continuously at the output shaft. It is normal to accept 10% overload capability for one hour in a 12-hour period, but this should be confirmed by the manufacturers.

A standby rating is not recognised by the current British Standards but some manufacturers will provide special ratings dependent on the required duty.

The relevant British Standards covering the rating of machines are as follows:

— BS EN 60034-2: 1999: *Rotating electrical machines. Methods for determining losses and efficiency of rotating electrical machinery from tests (excluding machines for traction vehicles)*[7]

— BS 5000: *Rotating electrical machines of particular types or for particular applications*: Part 3: 1980 (1985): *Generators to be driven by reciprocating internal combustion engines*[8]

— BS ISO 3046-1: 2002: *Reciprocating internal combustion engines. Performance. Declarations of power, fuel and lubricating oil consumptions, and test methods. Additional requirements for engines for general use*[9].

It should be noted that the site conditions play an important part in rating a set. These include the operating ambient temperature, the relative humidity and altitude. In the case of the turbocharged reciprocating engine with charge air cooling and water-cooled intercooler the lowest good quality water temperature available is relevant to the site rating.

Typical values relating to site conditions for a standard set are:

— maximum ambient temperature of 40 °C

— maximum altitude of 1000 m

— harmonic content not greater than 10%

— power factor of 0.8 lagging.

Full information on the values can be found in BS EN 60034-2[7] and BS 5000: Part 3[8].

Detailed information giving manufacturers' guaranteed output for the prime mover and alternator are given on the rating plate of the set. This provides a permanent record.

*British Wind Energy Association, 4 Hamilton Place, London W1V 0BQ (www.bwea.com)

4.2.2.6 Prime mover starting

The prime mover is started by means of an electric motor or direct injection of air into cylinders. The latter applies more usually to slow running sets. High speed sets of up to 1800 kV·A for standby use are commonly started using an electrically powered starter motor.

Provision is made for multiple starts, typically three, on a single occasion and the battery or air supply is sized accordingly. After three failures to start, the set would normally be shut down and an alarm sounded.

4.2.3 Generator size

4.2.3.1 Generator size related to load

When considering private or standby power plant it is necessary to carry out a full investigation into the characteristics of the load that has to be catered for, such as the method and type of starting of the largest motor, the number of motors likely to be starting together, the existence of non-linear loads etc. The end result of such an investigation will ensure that the units installed will meet the load pattern and total power required.

4.2.3.2 Load characteristics

The different load characteristics likely to be encountered will fall under one or more of the following:

— electric motors for normal continuous use, such as driving fans and pumps for building services and motor drives in process and industrial plant

— electric motors for intermittent operation, such as lifts, multi-stage refrigeration equipment, cranes and process plant needing high inertia starting

— non-linear loads, including discharge and fluorescent lighting and solid state controlled devices for battery charging, uninterruptible power supplies and rectifiers and switch mode power supplies on variable speed motor drives

— large linear loads, including electric heating and hot water systems, and tungsten lighting (unless fitted with thyristor-controlled dimmers)

— industrial welding equipment

— loads requiring stability of voltage, frequency and waveform within specified tolerances to ensure correct operation; these include mainframe computers, X-ray machines, and other medical diagnostic equipment.

In most buildings a mixture of the types of load mentioned will normally be found. It is necessary to consider the relative properties and characteristics of these load to ensure the proper selection of generating equipment.

4.2.3.3 Choice of engine and alternator

The rating of engine and alternator needed properly to support the load under start and operating conditions should be carefully considered. An oversized system is expensive and this includes the space it takes up and associated distribution system. Furthermore it will not operate at an efficient point on its load curve and, in the case of a diesel prime mover at low loads, could fail due to over-oiling.

4.2.3.4   Loads and characteristics

General motor loads

Industrial and commercial installations normally include induction motors (i.e. machine drives, fan drives, compressors etc.).

A substantial voltage dip may occur at the alternator terminals during motor starting, affecting the starting performance, and other connected loads. This can be estimated by referring to the nomograms and data provided by the manufacturers.

The rated kV·A of an alternator is related to the output rated voltage and full load running current. The total apparent power for a three-phase alternator is given by:

$$S_a = \sqrt{\frac{3 V_1 I_1}{1000}} \qquad (3.1)$$

where S_a is the rated kV·A of the alternator (kV·A), V_1 is the rated output line voltage (V), and I_1 is the rated full load line current (A).

The starting kV·A of the motor is related to the voltage and current at the moment of switch-on of the motor, and should be considered when selecting the size of the alternator.

Regenerative effects

A generator set providing power for motor drives with dynamic or regenerative braking (e.g. cranes, lifts) need to be capable of accepting the regenerated power, and reverse power protection must be applied to prevent 'motoring' that could result in damage to the alternator and, possibly, the prime mover. Special starters are available that will prevent regenerated feedback.

Non-linear loads

Solid state devices which include discharge and fluorescent lighting, rectifier/invertors and computer power supplies, do not draw sinusoidal currents from the supply. The effect can be a distorted current wave which may contain a high percentage of third harmonic (150 Hz) current (up to 80% of the total current). The result is a flow of third harmonic current within the generator windings which could cause overheating of the windings and connecting cables.

The line conductors should be protected by conventional over-current devices; however the neutral is not so protected. For sensibly balanced 50 Hz loads the neutral current will be relatively low. Where triple *n* harmonics exist (i.e. third harmonic plus multiples), the possibility of overloading the neutral conductor must be carefully considered in consultation with the equipment manufacturer.

4.2.4   Choice of high or low voltage systems

4.2.4.1   Cost implications

Throughout Europe, the normal LV distribution is 400/230 V three-phase, four-wire at 50 Hz, with a tolerance of +10% and –6%. The generation voltage will normally be at one of the distribution voltages already in use on the site. As a result, if high voltage is used, it will generally be due to the fact that this is already available from the REC with metering.

Justification for high voltage generation will be made using assessments similar to those for any other electrical distribution systems.

4.2.4.2   Operational and other implications

All electrical installations require stringent safety considerations, irrespective of the voltage at which they operate, and are subject to the Electricity at Work Regulations 1989[10]. Only authorised and properly trained personnel should be involved in the operation of generation plant and no work should be carried out without certification that the equipment has been isolated, locked off and earthed. A safe-working installation has to be provided with barriers, together with a description and sketch on the 'authority to work' permit.

It is possible to utilise outside qualified organisations, such as the RECs, to undertake switching and maintenance operations.

In some circumstances it may be more attractive to use a number of smaller low voltage sets around the site to eliminate expensive long cable runs, which may provide a greater, desired level of security at the individual centres.

High voltage generation is more cost effective above about 1.25 MV·A. These sets are normally manufactured for a specific project with resultant longer delivery and little chance of ex-stock sets.

The fault level increases with the rating of the set up to a point where it becomes necessary to review the earthing of the high voltage system but this can be reduced by introducing an earthing transformer, reactor or resistor.

In these circumstances it is recommended that a specialist be consulted.

4.2.5   Mode of operation

4.2.5.1   General

Generator systems may be provided for standby, peak-lopping or base load with operation modes outlined below.

In a standby application the system may consist of one or more sets to cover all or part of the normal load requirements, but in some applications a redundant set could be required for increased security.

Peak-lopping is often economical where an abnormally high load occurs at certain known times each day. Generating sets may be installed to cover these daily load peaks, and such sets, which should be continually rated, are then available as standby sets in the event of mains failure. Peak-lopping will reduce the registered maximum demand on the REC metering and hence reduce the tariff charge.

In base load applications there may be an option to obtain a standby facility from the REC but this may not be economically viable. It may be necessary to provide a redundant set to cover for maintenance and failure.

4.2.5.2   Individual generating systems

If the individual set required for standby is well maintained and used at times within the running programme of the station, there is little likelihood of failure.

4.2.5.3   Parallel operating generator systems

If there is a need to match load efficiency, reliability and increase flexibility, together with allowing one set to be off-line for maintenance, parallel operation of the system will meet one or more of these requirements.

Operating efficiency and running costs are affected by the management of the load in relation to the number of sets in operation.

Automatic control systems are available which provide continuous load optimisation while the plant is running. As the load is reduced, sets are disconnected progressively and shut down. Such control systems usually enable the operating parameters to be adjusted to avoid unnecessary operations and to maintain a minimum percentage of spinning reserve (i.e. the spare generating capacity of sets running on load). This allows for starting of larger loads and for fault clearing.

It is recommended that a specialist be consulted where there are sets operating in parallel, or in parallel with the REC mains.

4.2.6   Installation

4.2.6.1   Space requirements and room layout

Generating sets are normally installed in purpose-built rooms, existing rooms, weatherproof enclosures or sound- and weather-proof enclosures. An enclosure may allow access to the set by either incorporating space on the inside, or by removable covers.

In hot climates the set may be located externally beneath a sun shade. The size of the space for the set and the access needed will increase with the rating of the engine. High speed engines tend to be more compact than low speed ones. The set itself comprises the prime mover (engine) and alternator. Allowance must be made for associated fittings and equipment.

Table 4.1 Minimum clear room dimensions for single generating sets of various ratings

Output / kV·A	Length / mm	Width / mm	Height / mm
30	4000	2500	2500
50	4350	2500	2500
100	6150	3500	3000
150	6150	3500	3500
250	6800	4500	4000
300	7100	4500	4000
400	7500	4500	4000
700	9000	5000	4500
1000	9000	5000	4500
1500	9750	6000	4500
2000	12 000	6500	5000

Typical examples of fittings and equipment include the set control panel, output switch, starting system, fuel storage, exhaust with silencer plus inlet, and exhaust cooling air with attenuators. The cooling of the diesel engine block is commonly by means of a water jacket and air-cooled radiator.

Minimum clear room dimensions for a range of ratings of single generating sets with a skid-mounted control panel and daily service tank are given in Table 4.1.

A gas turbine prime mover is much smaller but demands considerably more combustion air than a reciprocating diesel engine. It also mixes the burnt gases with the outlet cooling air prior to exhausting them. The turbine would require larger attenuators. It is fully air-cooled.

4.2.6.2   Cooling

Where a diesel-engined generating set is proposed, cooling will usually be by means of a set-mounted radiator with fan assisted airflow for exhaust air, and supply air entering the generator room naturally through wall mounted louvres. However, there are alternative ways of achieving cooling.

Where space is limited in the area available for the set, or there is inadequate access to outside air, then the radiator for a water-cooled engine may be remote from the engine. If the distance is only a couple of metres, then the existing or uprated water pump may be adequate with a piped connection.

If the radiator is removed to another floor of a building or is more than a few metres from the engine, then a water-to-water heat exchanger and pumped secondary water circuit may be necessary to minimise operating pressures on the water jacket. Consideration must be given to maintaining reliability and providing desirable engine operating temperatures.

Some diesel engines with exhaust driven turbochargers may require cooling for the charge air intercooler. This will need to be accounted for in any modification to maintain the rating of the prime mover.

The advice of the set supplier and engine manufacturer should be sought.

4.2.6.3   Environmental requirements

The environment should be considered both for its effect on the set and the effect of the set on it. The prime mover will demand air for combustion. As a result of combustion, it will discharge noxious fumes and noise into the environment.

Air at high altitudes is less dense than at sea level. This will reduce the power output of the prime mover, affecting a turbine more than a diesel engine.

A turbine will be de-rated as air temperature rises because the maximum power output is governed by the combustion temperature. This occurs to a lesser extent with a diesel engine.

For anything other than normal temperatures and altitudes, the advice of the engine manufacturer should be sought to ensure that the prime mover will deliver adequate power to the alternator.

Atmospheric humidity is less of a factor when considering engine performance, but it could lead to corrosion.

It is common with standby sets to provide some means of warming the engine lubricant to ease starting, reduce wear at the moment of starting and assist rapid load acceptance. Where necessary, manufacturers can provide a lubrication oil priming pump with timed operation to maintain an oil film around the bearing surfaces.

There is little point in fitting highly effective silencing if the room itself does not by virtue of its mass provide a similar level of sound reduction. In lightweight structures used today, it may prove necessary to introduce denser materials for the room containing the set. Doors must be selected to effect adequate sound reduction and be a good fit to avoid leakage of noise.

Alternatively, the set may be housed in a sound-proof enclosure within the building.

To avoid transmission of noise and vibration through the structure, the set should be supported on anti-vibration mountings. Associated equipment connected to the set and to the structure should be fitted with flexible joints.

4.2.6.4   Primary fuel considerations

The primary fuel to the prime mover may be natural gas, liquefied petroleum gas, petroleum, diesel or industrial fuel oil.

Diesel oil is commonly used to fuel standby sets. It is readily available, easy to store and relatively inert. As the amount of fuel used is low, the cost of the fuel is less relevant than the capital cost of the set and its installation.

For base load and continuously operated sets, the type, availability and cost of fuel becomes a significant factor. The fuel may be available as a factory or process by-product, e.g. steam, where it is desirable to convert it to electrical energy. There are numerous options and alternatives which are outside the scope of this section of the Guide.

Diesel oil systems

It is usual to provide a day tank fitted above the set or incorporated into the base frame.

Diesel fuel may be stored in bulk in surface, buried or underground tanks. This allows fuel to be pumped to the day tank and returned to the bulk tank in the event of an overflow. The pumping would be controlled by level switches. The day tank provides an acceptable source for the engine fuel pump.

Tanks may be buried or placed in bunded pits to fully contain any spillage. The local authority may insist on a particular arrangement.

A fill point for delivery tankers should be provided within reach of their hoses. These are generally in lengths of 15 m and 30 m. In special circumstances, longer hoses can be supplied.

The fill point will typically contain the tanker connection point, a level indicator and a 'tank full' alarm depending upon requirements.

Typical dimensions for cylindrical tanks for a range of capacities are listed in Table 4.2.

Table 4.2 Typical dimensions for cylindrical tanks of various capacities

Capacity / litre	Diameter / m	Length / m
5000	1.50	2.80
10 000	2.00	3.00
25 000	2.50	5.00
50 000	2.75	9.00

Stored energy systems

In appropriate circumstances, the source of energy may not be a fuel, it may be a form of stored energy. This would include water pumped to a higher level reservoir to provide a static head or air held under pressure, perhaps in underground caverns.

In these cases, the prime mover would comprise a water or air turbine. The same turbine driven in reverse may be the means of re-establishing the stored energy. Alternatively a completely independent pump of some type may be used. There is a good example of pumped water storage in Dinorwic in North Wales.

4.2.7   Testing and commissioning

4.2.7.1   Introduction

The testing and commissioning of a generating set (or combination of sets) and the associated control equipment are vital functions needing to be properly carried out and recorded. Tests will normally be divided into the following:

—   *works tests*: covering the manufacture, assembly and performance of the main component items of the generating plant and control equipment

—   *site tests*: proving the installation standards and operational performance, under site conditions, of the complete generating plant installation including interfaces with other associated engineering systems.

4.2.7.2   Works testing

General

The complete generator set assembly, including its control equipment and switchgear, must be tested as a composite unit at the generator set manufacturer's works in order to prove that its performance meets specified requirements, in so far as is practicable in the absence of the actual dynamic loads of the eventual building installation.

With large base load sets, however, it may often be impracticable to perform a full works test on the complete engine, alternator and auxiliary equipment assembly. Such sets are more usually subjected to works tests on the separate alternator and engine assemblies at the appropriate manufacturer's works, with combined assembly tests being carried out on-site as part of the extensive erection and testing procedures.

It is important to maximise the benefits gained from works tests rather than to leave certain tests to be carried out during on-site commissioning since the latter option is likely to create programme delays in completing the installation should a malfunction be found.

All reputable generator manufacturers have their own series of works tests and details of these should be sought from the particular manufacturer. A code of practice[11] published by the Association of British Generating Set Manufacturers (ABGSM) sets down a basic standard test schedule. However, the basic tests included in this schedule should be regarded as the minimum required and it may be appropriate to specify that additional 'type tests' also be carried out.

The basic requirements for performance are set down in BS EN 60034-2[7], BS 5000[8] and BS ISO 3046-1[9].

Standard tests

These should be conducted on every generating set before despatch and should include the following standard tests. The main components of this standard test schedule are listed in the preferred sequence:

(*a*)   Check plant build against specification.

(*b*)   Record reference and rating data.

(*c*)   Prepare generating set for test with connections for exhaust, cooling system, lubricating oil, electrical power, control wiring etc.

(*d*)   Preliminary starting and running checks/adjustments.

(*e*)   Performance tests (initial set up checks).

(*f*)   Load acceptance tests.

(*g*)   Governing and voltage regulation (cold conditions).

(*h*)   Load duration tests with detailed examination of set immediately following hot shutdown.

(*i*)   Governing and voltage regulation (hot conditions).

(*j*)   Transient switching tests.

(*k*)   Insulation tests.

(*l*)   Final re-run of set to observe correct performance.

(*m*)   Final check and clearance including draindown, disconnections and completion of test certificates.

Type tests

Apart from the above standard tests, there may be a requirement for certain type tests to be carried out at the manufacturer's works. The requirement to perform such tests must be agreed beforehand with the manufacturer and therefore should be included in the specification.

Examples of such tests include the following:

—   cold resistance of electrical windings

—   winding resistance and temperature rise following hot shutdown

—   transient performance test

—   vibration tests

—   noise tests

—   ability to withstand short circuit.

Any type test certificates required should be requested from the manufacturer.

4.2.7.3   Site testing and commissioning

The generator plant, including its associated fuel, cooling, exhaust, electrical and fire safety systems, must be thoroughly checked for correct operation in both manual and automatic modes during, and upon completion of, site installation. The initial checks will consist largely of static checks to ensure that the installation has been carried out correctly in readiness for full testing and commissioning.

The engine/alternator shaft alignment must be checked where the set is not skid-mounted or where the assembly has been dismantled for transportation to the site.

The generator output voltage, frequency and phase rotation should be checked before further tests are carried out and before final connection to the electrical distribution system.

It is desirable to carry out a load test with the standby generator supplying the full site load. Since it is unlikely that the actual building load will be available at the commissioning stage, consideration should be given to the inclusion in the contract of a suitably rated load bank. Such a load bank should be specified as multi-stage and preferably capable of offering loads at power factor 0.8 lagging. The test load bank may also provide a test facility for routine maintenance where it is not practicable or desirable to use the actual building load.

The duration of the site full load test should be at least four (preferably eight) hours for all sets above 100 kW. Sets rated at less than 100 kW are usually tested for one hour.

In all cases, the test should be carried out at varying loads up to the full rated load, with instrument readings recorded every fifteen minutes. These readings should include jacket water temperature, exhaust temperature, lubricating oil level, temperature and pressure, output voltage and frequency. The room temperature should also be monitored.

It may be necessary to use an oscilloscope for voltage and frequency measurements, particularly where close tolerances are specified.

Noise and vibration measurements should also be monitored if the generator installation is likely to cause a disturbance to nearby occupants.

Multiple set installations will need to be commissioned so as to demonstrate synchronisation with any combination of sets available and, where appropriate and by prior agreement, in parallel with the REC. In the latter case, it is often possible to arrange for power to be exported to the REC network for load testing purposes.

All local/remote controls, emergency stop and all protection devices/interlocks etc. should be proven during on-site commissioning tests.

The time interval should be noted between initial start up and load acceptance for the lead machine and, for multiple set installations, the time taken for the remaining sets to synchronise and accept load. Load sharing between sets should also be checked.

Site testing of a generator set intended to supply power and start a large induction motor, such as a chiller compressor, may prove to be impracticable at the time of commissioning the generator installation. However, it is vital that this is proven before final acceptance of the installation. In advance of this test, the generator works test data should be checked against the motor manufacturer's data for compatibility with the performance requirements. The motor data should include proven data on the characteristics of the motor starting device in starting the actual motor to similarly prove compatibility.

4.2.8 Operating and maintenance

Upon completion of all site commissioning tests the generating set supplier should provide the following:

— an appropriate period of user training before final handover to the building owner/operator

— full operating and maintenance instructions, together with comprehensive record drawings and descriptive documentation upon handover

— any special tools required for maintenance (irrespective of who will maintain the equipment)

— spare parts (as recommended and/or included in the contract price)

— proposals for ongoing maintenance contract (if required).

4.2.8.1 User training

It is vital that the staff responsible for operating the generating plant are given proper instruction prior to handover by the equipment supplier.

This must particularly deal with all safety aspects in accordance with the Health and Safety at Work etc. Act 2002[12], the Electricity at Work Regulations 1989[10], and any HSE guidance notes.

Apart from the moving parts of the equipment, the user's attention must also be drawn to the safety and health aspects of ancillary static components such as batteries, fuel oil, anti-freeze, lubricating oil and electrical hazards.

The period of user training will vary depending upon the number and size of sets, calibre/experience of the operating staff, any ongoing involvement from the generating set supplier etc., but as a preliminary guide, should be for a minimum of one full working day.

It is useful for the user to be taken through the commissioning reports, with demonstrations wherever practicable, as part of the training. This is in addition to a thorough explanation of the content of the full operating and maintenance instructions.

4.2.8.2 Operating and maintenance instructions

The operating and maintenance manual should cover all parts of the generating plant installation including the prime mover, alternator, control system and all ancillary equipment.

The contents should include:
— data sheets

— health and safety warnings

— description (general and detailed)

— installation guidance notes

— commissioning procedures and records/test sheets

— general arrangement drawings

— circuit/schematic diagrams

— operating instructions

— maintenance instructions

— fault finding guidance

— recommended spares and ordering information.

4.2.8.3 Special tools

The generator supplier must provide, upon handover, all special tools necessary to maintain or operate the equipment in a safe and proper manner.

These tools are necessary irrespective of whether or not the supplier has a maintenance contract for the installation.

Details of such tools should be included in the operating and maintenance manuals.

The tools should be housed in a clearly marked location, preferably close to the main generating plant equipment.

4.2.8.4 Spare parts

The contract specification for the generating plant should allow for the provision of those spare parts recommended by the supplier to be immediately available for maintenance purposes. Such spares should therefore be provided upon handover to the user.

Other spare parts, to be ordered as necessary, should be detailed in the spares section of the operating and maintenance manual. This should include all references and sources where these can be obtained, together with ordering information.

4.2.8.5 Maintenance contract

Serious consideration needs to be given at an early stage of the contract to the ongoing maintenance responsibilities.

The building owner/main tenant may already have suitably qualified and experienced engineering staff capable of carrying out the necessary maintenance activities.

A serious alternative would be to negotiate a maintenance contract with the generator set supplier. This may be preferred whether or not the building owner/main tenant has also own in-house engineering staff. It is likely to be particularly beneficial where there is a multi-set installation.

4.2.8.6 Periodic maintenance

Generating plant, in particular the engine system, requires regular attention to ensure reliability, optimum performance and long life.

All maintenance procedures should be fully in accordance with the generator set operating and maintenance instructions. A maintenance log book should be established upon accepting the plant after commissioning. This should include a full record of hours-run readings and a summary of all servicing and repairs carried out, together with fuel and lubricating oil consumption totals.

Whatever the mode of operation, the generating plant must operate when called upon to do so. Failure of the battery starting system will prevent the generator set starting and hence this needs regular and frequent checking.

Periodic maintenance recommendations from the supplier will normally cover routine checks on a frequent basis (possibly after 100 hours of use or at monthly intervals) depending upon various operating factors.

More detailed maintenance, including replacement of items such as lubricants, filters, belts and coolant, can be expected every, say, 500 hours.

Major maintenance activities are required typically after 5000 hours. This may involve a complete strip-down of the engine, and inspection and replacement of worn or defective parts.

References

1 Electricity Act 1989 Chapter 29 (London: Her Majesty's Stationery Office) (1989)

2 Electricity Safety, Quality and Continuity Regulations 2002 Statutory Instruments 2002 No. 2665 (London: The Stationery Office) (2002)

3 BS 7430: 1998: *Code of practice for earthing* (London: British Standards Institution) (1987)

4 *Recommendations for the connection of private generating plant to the Electricity Boards' generating system* Electricity Association Engineering Recommendation G59/1 (London: The Electricity Association) (1991)

5 *Notes of guidance for the protection of private generating sets up to 5 MW for operation in parallel with PES distribution networks* Electricity Association Engineering Technical Report ETR113 (London: The Electricity Association) (1995)

6 *Energy efficiency in buildings* CIBSE Guide F (London: Chartered Institution of Building Services Engineers) (2004)

7 BS EN 60034-2: 1999: *Rotating electrical machines. Methods for determining losses and efficiency of rotating electrical machinery from tests (excluding machines for traction vehicles)* (London: British Standards Institution) (1999)

8 BS 5000: *Rotating electrical machines of particular types or for particular applications*: Part 3: 1980 (1985): *Generators to be driven by reciprocating internal combustion engines* (London: British Standards Institution) (1985)

9 BS ISO 3046-1: 2002: *Reciprocating internal combustion engines. Performance. Declarations of power, fuel and lubricating oil consumptions, and test methods. Additional requirements for engines for general use* (London: British Standards Institution) (2002)

10 The Electricity at Work Regulations 1989 Statutory Instruments 1989 No. 635 (London: Her Majesty's Stationery Office) (1989)

11 *A code of practice for designers, installers and users of generating sets* ABGSM TM3 (Redhill: Association of British Generating Set Manufacturers) (1985)

12 The Health and Safety at Work etc. Act 2002 (London: Her Majesty's Stationery Office) (2002)

Bibliography

Electrical services: supply and distribution Health Technical Memorandum HTM 7 (London: Department of Health and Social Security/Welsh Office/HMSO) (1977)

Emergency electrical services Health Technical Memorandum HTM 11 (London: Department of Health and Social Security/Welsh Office/HMSO) (1974)

5 High voltage distribution

5.1 Introduction

The primary reason for the transmission of power at high voltage (HV) is an economic one. The power loss in a cable is a function of the square of the current carried, as given by the formula:

$$P_{L} = I^2 \times Z_{c} \qquad (5.1)$$

where P_{L} is the power loss (W), I is the current carried in the cable (A) and Z_{c} is the impedance of the cable (Ω). By increasing the voltage and, hence, reducing the current, the power loss is reduced and the transmission efficiency improved.

In this section, consideration is given to system design and to the selection and installation of equipment and cables. Consideration is also given to the calculations involved in assessing fault levels and the protection necessary to deal with over-current and earth faults.

5.2 Standards

The main standards that apply to HV switchgear and cables are as follows:

— BS EN 50187: *Gas-filled compartment for AC switchgear and controlgear*[1]

— BS EN 60129: *Specification for alternating current disconnectors and earthing switches*[2]

— BS EN 60255: *Electrical relays*[3]

— BS EN 60265: *Specification for HV switches*[4]

— BS EN 60282: *HV fuses. Current limiting fuses*[5]

— BS EN 60298: *AC metal-enclosed switchgear and controlgear*[6]

— BS EN 60420: *Specification for HV AC switch-fuse combinations*[7]

— BS EN 60470: *HV alternating current contactors and contactor-based motor starters*[8]

— BS EN 60644: *Specification for HV fuse-links for motor control applications*[9]

— BS EN 60694 : *Common specifications for HV switchgear and controlgear standards*[10]

— BS EN 61330: *HV/LV prefabricated sub-stations*[11]

— BS EN 62271: *HV switchgear and control gear*[12]

— BS 159: *Busbars and busbar connectors*[13]

— BS 923: *Guide on HV testing techniques*[14]

— BS EN 60044-1: *Instrument transformers. Current transformers*[15]

— BS 3941: *Voltage transformers*[16]

— BS 5207: *Specification for sulphur hexafluoride for electrical equipment*[17]

— BS 5311: *HV alternating current circuit breakers*[18]

— BS 5992 (IEC 60255): *Electrical relays*[19]

— BS 6480: *Impregnated paper-insulated lead or lead alloy sheathed cables of rated voltages up to 33 000 V*[20]

— BS 6553: *Guide to the selection of fuse links of HV fuses for transformer circuit applications*[21]

— BS 6622: *Cables with extruded cross-linked polyethylene or ethylene propylene rubber insulation for rated voltages up to 19/33 kV*[22]

— BS 6626: *Maintenance of electrical switchgear and controlgear for voltages above 650 V and up to and including 36 kV*[23]

— BS 6878: *HV switchgear and controlgear for industrial use. Cast aluminium alloy enclosures for gas-filled HV switchgear and controlgear*[24]

— BS 7197: *Performance of bonds for electric power cable terminations and joints for system voltages up to 36 kV*[25]

— BS 7315: *Wrought aluminium and aluminium alloy enclosures for gas-filled HV switchgear and controlgear*[26]

— BS 7835: *Armoured cables with extruded cross-linked polyethylene or ethylene propylene rubber insulation for rated voltages up to 19/33 kV having low emission of smoke and corrosive gases when affected by fire*[27].

5.3 Typical operating voltages

Electrical supplies are generally distributed at a number of standard voltages, typically 400 kV, 275 kV, 132 kV, 33 kV and 11 kV. This is discussed in section 4.1.

However, in some parts of the country other voltages may be provided by the Regional Electricity Company (REC) e.g. 20 kV, 6.6 kV, 3.3 kV. The reasons for this are mainly historical due to older equipment (particularly transformers) being installed and the cost of replacement to provide a standard voltage being prohibitive.

In general, the building services engineer is unlikely to be involved with voltages greater than 11 kV.

The selection of the most appropriate voltage will depend upon:

(*a*)   the voltage available from the REC

(b)   the load capacity requirement

(c)   the general nature of the load and load profile i.e:

—   constant

—   motor starting

—   harmonic generation.

5.4   HV system design

An HV distribution system is used in preference to an LV distribution system because of the following advantages:

(a)   For the same electrical load the current requirement is significantly reduced. Typically, 1000 kV·A of AC power at 400 V, 3-phase equates to a balanced load of 1443 A per phase but only 52 A at 11 kV.

(b)   Because of (a) above, it is possible to use smaller diameter, lower cost, cables when power is transmitted at higher voltages. Typically, a balanced load of 1000 A would require two 240 mm^2 copper conductors per phase at 400 V (assuming buried, XLPE-insulated, multi-core cables) but only a single 25 mm^2 (the minimum standard size available) 3-core cable at 11 kV.

(c)   The voltage drop along the distribution system is reduced and is a smaller fraction of the initial voltage

(d)   The power losses in the transmission system are reduced.

Disadvantages of using HV as a distribution system are:

(a)   increased danger because of the higher voltage

(b)   the need for a higher level of insulation, which is more expensive

(c)   higher levels of training required to install and maintain the equipment.

The first step in designing a distribution system is to establish the magnitudes and locations of loads. Having established these, the designer should identify the number and locations of switchrooms and substations. Since LV distribution is generally more expensive (per kV·A carried) than HV distribution, and problems with voltage drop appear if the LV cable runs become too long, it is often better to have too many substations than too few.

5.4.1   Substations

5.4.1.1   Types of substation

If an HV distribution network is being used then it will consist of the following types of substations.

—   *Intake substation*: This substation is the point of incoming supply from the electricity supply company.

—   *Distribution substation*: This substation is the point of distributing the electrical energy to the load centres.

—   *Standby substation*: This substation is provided to control standby generators if they feed energy into the HV distribution network.

The relative functions of these types of substations is shown in Figure 5.1. Different types of substation may be combined in a single location.

5.4.1.2   Location of substations

Intake substation

If the intake substation is to be the only substation on site then it should be placed as near to the load centre as possible. This is because LV cabling is much more expensive than HV cabling. Other considerations are:

—   the point of supply of the electricity supply company (usually the Regional Electricity Company (REC))

—   access requirements

—   any physical restraints.

If the intake substation is part of a large network with many distribution substations, then its location is of less importance. In this case it should be sited at a point for easy access for the point of supply from the electricity supply company.

Distribution substation

The ideal positions for distribution substations are at the load centres that they serve.

Standby substation

The standby substation has the control circuit breakers for the standby generator sets and should therefore be sited as near to the standby generators as possible.

5.4.2   Distribution methods

Where HV distribution is used within a site, various arrangements are possible. What is best for any particular project will be influenced by the magnitude, location and importance of the loads. The eventual choice will be an economic one, based on the costs to the consumer of the loss of electrical supply to a particular area, weighed against the cost for the provision of the increased redundant electrical capacity.

The design of the system should be kept as simple as possible; simple and straightforward systems are easier to operate and less expensive to maintain.

The design should also consider the expected expansion of the site. If the area is restricted, future development is unlikely and the HV distribution system provided will be on a 'once and for all' basis. However, if areas are available within the site for further development, the long term plans for these areas should be ascertained and the HV system configured to accommodate possible future expansion. A decision taken at this stage will affect the site for some 20 or 30 years to come on the basis that the HV

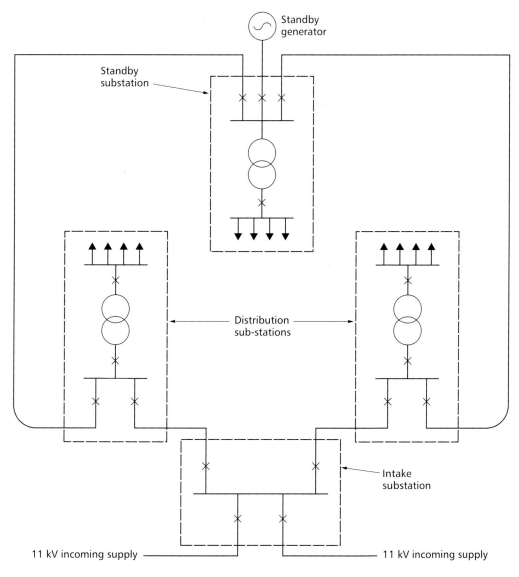

Figure 5.1 Schematic arrangements of substations serving an industrial site to illustrate different substation types

distribution system is a long-term infrastructure provision.

There are two basic methods of distributing loads on HV systems. These are 'radial feeders' and 'ring mains'. These are illustrated in Figures 5.2 and 5.3 respectively. In each case, it is assumed that the local REC will provide a 3-phase supply at 11 kV and terminate this in HV switchgear that is part of the REC's network.

5.4.2.1   Radial feeders

The arrangement shown in Figure 5.2 is the one that the building services engineer is most likely to encounter in a typical project where the REC provides the substation. The simple radial system shown consists of only one substation, which incorporates both the incoming supply and the distribution switchgear. In this arrangement, the REC switchgear is usually a ring main unit (as their supply is likely to be taken from an HV ring circuit) incorporating isolating switches on the ring main and a fuse switch or circuit breaker for the consumer. If the ring main unit is located adjacent to the transformer, no additional HV isolation is required. However, if the transformer is located at a distance from the supply switchgear, a local isolating switch should be installed at the transformer.

The arrangement shown in Figure 5.2 is suitable for installations which the supply serves only one building or

where security of supply is not particularly important. As the size of the site or building increases or where the load increases, it will be necessary to provide additional substations as radial supplies.

Radial circuits are the easiest to design and cheapest to install, but they do not provide the provision of alternative methods of supply that ring mains can offer.

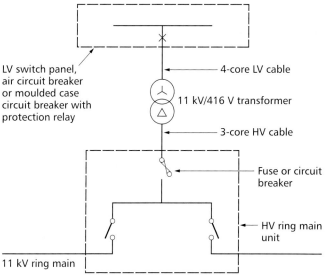

Figure 5.2 Radial HV distribution system

The type of switchgear used for this type of distribution will dependent on the size of the load that is being supplied. If the load is less than 1000 kV·A and there is no standby generation then the normal method of supply is by fuse switches or ring main units because of their cost advantage over circuit breakers.

If the load of the distribution centres is greater than 1000 kV·A and if there is standby generation then the switchgear required will usually be circuit breakers.

5.4.2.2   Ring mains

Where more than two substations are installed on radial circuits, consideration should be give to connecting the substations into a ring circuit. This provides the user with the ability to isolate items of switchgear or sections of the circuit for maintenance without losing the supply to all the loads.

There are two types of ring mains, open or closed. The system that is chosen will depend on the level of security that the distribution system is being designed for.

Open ring main

Figure 5.3 illustrates an open ring main using ring main units.

With an open ring system, all of the substations are connected together in a ring configuration but one of the switches in the ring is left open. Effectively, the ring

circuit becomes two radial circuits but it permits simple reconfiguration of each side of the ring by changing the open point of the circuit.

This type of network, when compared with a radial service, gives improved speed of restoration of service in the event of a fault and more flexibility when carrying out maintenance.

When selecting the location of the open point of the ring the following points need to be considered:

—   the importance of the individual distribution substations

—   the load on each side of the open ring

—   the cable length involved between distribution substations.

If there were to be a fault on one of the cables in the ring, the supply circuit breaker serving that side of the ring would trip. This would result in the loss of supply to all the distribution substations between the circuit breaker and the open point in the ring. When the location of the fault is identified, the faulty part of the ring main can be isolated and, by moving the open point of the circuit, the other substations re-energised.

Closed ring main

A closed ring main is illustrated in Figure 5.4. In this system, substations use circuit breakers for protection and control.

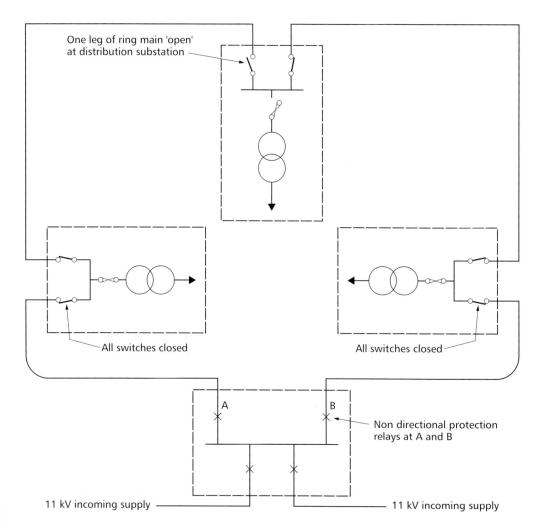

Figure 5.3 Open ring main HV distribution system

A closed ring main is where the intake and distribution substations are connected in a ring configuration and all the switchgear is closed. However, as each substation is fed from both directions of the ring main, it is necessary to provide sophisticated unit protection against over-current and fault currents that can flow in either direction around the ring main, depending on the location of the fault. This is discussed in more detail in section 5.6.7

With the closed ring main, all the switchgear used will be circuit breakers. Under normal conditions they will all be 'on', unless they have interlocking devices (standby breakers). On each of the circuit breakers, network protection equipment will be fitted to isolate any faulty piece of apparatus in the event of a fault. For example, if there were a fault in one of the interconnecting cables then the two connecting circuit breakers at either end would be automatically disconnected, isolating the faulty piece of equipment only.

The network protection system (unit protection) on each of the circuit breakers is interconnected by a signalling cable between the substations.

Switching of HV systems must only be carried out by HV-authorised persons (see section 5.7.1) who have the necessary training and experience to safely operate these systems. In view of the cost of providing such personnel, this function is often contracted to a specialist HV contractor or to the local REC.

5.4.3 Design of substations

The design, layout and size of a substation is primarily determined by the type and extent of equipment that is being installed. Manufacturers' dimensional data should be consulted to ensure that adequate space is provided.

The cubicle switchgear in an indoor substation will normally be sited in a panel. The minimum clearance around this panel should be 800 mm. However, it is normal to have a clearance of 2000 mm at the front of the panel to allow the truck part of circuit breakers to be withdrawn and maintained, and for access in front of the circuit breakers with the panel doors open and the circuit breaker trucks rolled out.

It is also normal practice for HV cables to be installed in trenches, rising into the back of the HV switchgear. Cubicle HV switchgear is generally designed for this installation method although the switchgear manufacturer can usually provide other cable arrangements for switchgear located in a basement where cables are installed at high level.

The minimum width of a cable trench should be 600 mm. The depth of the trench should depend on the finished ground level outside and inside the building but it should be deep enough to cater for the bending radius of the cables. The trench should be protected by covers made of metal or hardwood. Hardwood covers have the advantage that, if they were to fall into the trench, they would do less damage to the cables than steel covers.

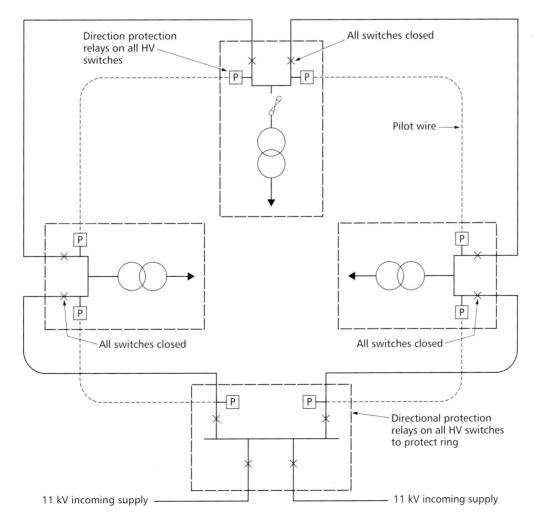

Figure 5.4 Closed ring main HV distribution system

Where an oil filled transformer is specified (see section 6), precautions should be included against the possibility of an oil leak. This should include appropriate bunded areas and provision for safely pumping any spilt oil.

Typical layouts for distribution substations are shown in section 7, Figure 7.7.

Where space is limited, it may not be appropriate to separate the HV switchgear, transformer and LV switchgear but rather to combine them in a composite unit called a packaged substation. In this arrangement, the HV and LV switchgear are connected directly to the terminations of the transformer using short busbar connections instead of cables, the whole installation being installed within a common enclosure. Packaged substations should comply with BS EN 61330[11].

5.4.2.3 Environmental considerations

Indoor substation

To ensure that the design of an indoor substation is adequate the following items need to be considered:

— The floor should be smooth and level and it should be so constructed that it can support the total weight of the equipment being installed (including the provision for future expansion). Where appropriate, a structural engineer should be consulted.

— Adequate access should be provided for installing and removing the largest transportable item of equipment.

— Lifting beams should be provided.

— It is not advisable to install a window in a substation as it could aid unauthorised access. Ventilation should be provided by louvres installed in the walls or doors.

— The ends of all cable ducts should be sealed after the cables have been installed. Where ducts enter buildings the seal should form a barrier against oil, gas, water, fire and attack by vermin.

— External doors should preferably open outwards to a free access area of hard standing.

— A substation should have an emergency exit which should be secured by panic bolts only.

— Each substation should have general electrical services, consisting of:

  (*a*)   general and emergency lighting

  (*b*)   heating

  (*c*)   ventilation

  (*d*)   a main earth bar which has the substation earth and the HV earth connected

  (*e*)   general socket outlets.

— In the substation the ambient temperature should be maintained between 10 °C and 35 °C. This is important if a transformer is sited in the substation because of the large amount of heat that it can dissipate.

— Safety signage should be provided in accordance with statutory regulations.

Outdoor substation

To ensure that the design of an outdoor substation is adequate the following items need to be considered:

— A concrete slab should be laid to support the weight of the equipment. The slab should be smooth and level. Around the slabs should be open pits with cable access to them. The pits should be of sufficient size to completely absorb the total volume of oil in any switchgear or transformers. After the substation has been installed and cabled the pits should be filled with 10–20 mm grade shingle. Space around the transformer and switchgear should permit the removal of the shingle in the event of an oil spill.

— The substation should have a surrounding fence or wall. This is a requirement of the Electricity, Safety, Quality and Continuity Regulations 2002[28] for all new substations.

— External power sockets should be installed to aid maintenance personnel in the use of lighting or power tools. The sockets should be in a lockable enclosure.

5.5   HV switchgear

HV switchgear is provided to:

— enable an HV network to be broken down into smaller sections for maintenance

— provide circuit protection

— isolate faulty pieces of apparatus.

The basic types of HV switchgear are:

— circuit breakers

— oil switches

— switch fuses (or fuse switches)

— ring main units

— isolators.

5.5.1   Circuit breakers

A circuit breaker is a switching device which is capable of making, carrying and breaking currents under normal circuit conditions and also making, carrying for a specified time and breaking currents under specified abnormal conditions such as a short circuit.

The standards for circuit breakers are defined in BS 5311[18]. The main differences between different manufacturers of circuit breakers is the type of arc suppression. On making or breaking a circuit an arc will be drawn between the fixed and moving parts of the circuit breaker. The size and duration of the arc is determined by the medium within which the switching occurs.

At one time, oil was the predominant insulation medium in circuit breakers due to its low cost and high availability and a considerable number of oil-filled circuit breakers continue in use throughout the country. In the 1970s, oil was overtaken in popularity by vacuum and sulphur hexafluoride (SF_6) switchgear as the cost of this equipment fell and the electricity supply industry embraced this new technology. However, SF_6 has been found to be ozone-depleting and considerable care should be taken in its specification and use.

A circuit breaker can carry out the following duty functions, but it can only carry a fault current for a specified time:

—   make 'normal' current

—   carry 'normal' current

—   break 'normal' current

—   make 'fault' current

—   carry 'fault' current

—   break 'fault' current.

5.5.2   Oil switches

An oil switch is a device that is suitable for making a circuit in oil under normal and fault conditions and is also capable of breaking a circuit in oil under normal conditions. This device can perform the same duties as a circuit breaker except for being unable to break the circuit under fault conditions.

Switches are fitted with earthing facilities to enable the cable side to be earthed. The arrangement is integral and the earthing switch has the same rating as the main contacts of the switch.

The switches normally have a mechanical arrangement on the operating handle to prevent the switch being directly switched on and then off immediately.

5.5.3   Oil switch fuses (or fuse switches)

These devices are the same as oil switches with the addition of HV high rupturing capacity (HRC) fuses to give a fault breaking capacity. They are normally used for the protection of distribution transformers up to 1000 kV·A.

Oil switches and switch fuses are extendible and can be connected together to form any combination required for a distribution substation.

5.5.4   Ring main units

This is a standard unit which consists of incoming and outgoing oil switches supplying a common busbar, and one or two outgoing fuse switches to provide power to dedicated loads.

Ring main units provide a very economic solution to providing HV switchgear in substations and are frequently used by RECs. They are suitable for external use, thereby often avoiding the cost of a substation enclosure.

However, ring main units are generally unable to accommodate system protection relays and are therefore limited in their ability to provide sophisticated system protection.

5.5.5   Isolators (disconnectors)

An isolator is a device that is capable of opening or closing a circuit either when negligible current is broken or made, or when no significant change in the voltage across the terminals of each of the poles of the isolator occurs. The device has no making or breaking capacity. The current standard for isolators and earthing switches is BS EN 60129[2].

5.5.6   Ancillary equipment

HV switchgear can be provided with a variety of ancillary equipment and features as described in the following sections.

5.5.6.1   Auxiliary switches

These are switches that work in conjunction with the main phase contacts of an HV switch to control a circuit for operating auxiliary devices. The switches are low current and low voltage devices which are operated by the same mechanism as the main switch.

5.5.6.2   Earthing

HV switchgear is usually provided with facilities to enable either the main busbar or the outgoing circuit to be connected to earth for the purpose of making either safe. The methods used for earthing are as follows:

—   Use of a specially designed earthing truck which replaces the circuit breaker and is located into position and closed to earth either the main busbar or outgoing circuit.

—   Fitting of special earthing gear. These are purpose made items which are first connected to a suitable earth point and then connected to the main busbar or circuit spouts.

—   Integral earthing where the HV switch itself is moved to an alternative position within its housing to connect to earth either the main busbar or the outgoing circuit as required.

5.5.6.3   Auxiliary transformers

HV switchgear can be fitted with protective devices and metering etc. The supplies for these are provided from voltage or current transformers fitted to the switchgear.

Voltage transformers (VTs) have a secondary voltage of 110 V and the primary voltage to suit the supply voltage. The transformers will normally be oil-immersed, with star wound HV and LV, both windings being fused. On the LV side it is normal for the yellow phase to have a linked connection to earth and the neutral fused. The standard for VTs is BS 3941[16].

Current transformers (CTs) normally have a secondary rating of 1 A or 5 A and the primary rating dependent on the load. Separate CTs are provided for protection and measurement, the protection CTs having a higher accuracy rating. The standard for current transformers is BS EN 60044-1[15].

5.5.6.4 System protection

Over-current and earth fault protection are usually provided by either electronic relays or time-limit fuse links operating trip coils in the circuit breaker. Historically, electromechanical relays were used to provide this protection. Whilst many electromechanical relays are still in service, their use has generally been superseded by electronic equivalents.

Circuit breakers are designed to accommodate system protection relays and are therefore used by system designers where distance protection or complex local protection is required. Ring main units are normally fitted with simple protection on the outgoing circuit, based on a time delayed fuse.

System protection is dealt with in more detail in section 5.6.

5.5.7. HV switchgear selection

5.5.7.1 Circuit breakers

The rating of circuit breakers is specified in BS 5311[18] and is given in the following terms:

(a) *Service voltage*: the upper limit of the highest voltage of the systems for which the circuit breaker is intended. The BS standard ratings are 3.6 kV, 7.2 kV, 12 kV and 13.8 kV.

(b) *Insulation level*: the maximum voltage that the switchgear can withstand for a short time under fault level.

(c) *Frequency*: normally 50 Hz.

(d) *Normal current*: the RMS value of the current which the circuit breaker shall be able to carry continuously without deterioration under the prescribed conditions. The normal BS ratings are 400 A, 500 A, 630 A, 800 A, 1250 A.

(e) *Short circuit breaking current*: the current that the circuit breaker can break at the stated voltage. The breaking capacity determines the maximum fault current that the circuit breaker can break. Under fault conditions the current that flows in a distribution system is made up of two components:

— a symmetrical AC component: the level of this is determined by the amount of power available and the network impedance

— a DC component: this is caused by the discharge of the magnetic energy into the fault and will be of short duration.

It is not easy to determine the value of the DC component because it depends on the value of the instantaneous voltage at the time of the fault. Also by the time a circuit breaker operates under fault

conditions, the DC current value of the fault would have decayed away. Therefore when considering the breaking capacity of a circuit breaker it is normal to consider only the AC component.

Therefore the rated short circuit current (RSCC) is equal to the RMS value of the AC component of the rated short circuit breaking current. The standard ratings are 6.3 kA, 8 kA, 10 kA, 12.5 kA, 16 kA, 20 kA etc. It is essential that HV switchgear is selected with the correct RSCC to ensure that no mechanical damage occurs in the event of a fault.

(f) *Rated short time making current*: the peak value of the maximum current (including DC) that the circuit breaker is capable of making and carrying instantaneously at the rated supply voltage. BS 5311[18] requires that the rated short time making current of a circuit breaker shall be the rated short circuit breaking current multiplied by 2.5.

(g) *Duration of short current*: the period of time for which the circuit breaker, when closed, can carry a current equal to its rated short circuit breaking current. These are normally expressed in kA for one second.

(h) *Closing mechanisms*: with a circuit breaker there are four types of closing mechanisms; these are classified by the method of achieving the power required to complete the closing operation. The four types are as follows:

— *Dependent manual* or *hand closing*: this is a closing operation by hand without the use of any stored or other supply of energy. The speed of closure is therefore dependant on the speed of the operator.

This type of circuit breaker is no longer manufactured although they are frequently found on existing systems. Due to the possible destructive consequences of the contacts being closed/opened slowly, warnings to that effect should be fitted onto the equipment and consideration should be given to the replacement of the equipment.

— *Independent manual closing*: this is a closing operation by hand in a single operation in which energy supplied during the initial part of the operation is used to close the circuit breaker independent of the operator. The energy is usually that of a compressed spring.

— *Spring assisted*: this is a form of closing by hand where springs are compressed by one operation and their energy is released by a separate later operation to close the contacts. This is one of the commonest types of circuit breaker found on HV distribution systems.

— *Power closing*: this is where the circuit breaker is fitted with a solenoid closing mechanism. Closure of the circuit breaker can be remote. This type of circuit breaker is commonly used in large installations such as industrial and defence establishments.

5.5.7.2   Switches

BS EN 60265[4] gives details for the service, definitions, design and construction, type and testing of HV switches.

Normally the parameters used in selecting switches are:

— service voltage

— normal current

— number of poles, normally three

— frequency, normally 50 Hz

— breaking current: the current that exists at the instant of contact separation expressed as an RMS value

— making capacity: the current that a switch is capable of making and carrying instantaneously at the rated voltage.

5.5.7.3   Fuse switches, switch fuses and ring main units

These are selected on the same basis as switches except for the addition of the HRC fuse.

5.6   System protection

5.6.1   Need for system protection

When a fault occurs in any electrical system, the energy that flows as a consequence usually has the potential to damage equipment and cables. In certain circumstances, it can also endanger life, either through electrocution or by consequential fire.

To minimise the damage and disruption caused by electrical faults, protection must be provided that will isolate that part of the system in which the fault has occurred. This protection can be provided by various types of fuses and relays, as discussed later in this section. However, before protection equipment can be selected, it is necessary to carry out an assessment of the anticipated fault levels in the network.

Advances in computer-aided design have produced excellent software for the calculation of fault currents and the selection of protective devices. However, it is important to understand the underlying principles prior to adopting a software-based approach.

5.6.2   Fault calculations

The fault level should be determined at various locations within the electrical network to ensure that the switchgear and cable which are provided can withstand the energy of a fault without damage or degradation.

It is usual to express the fault level in HV and LV networks in terms of MV·A at the supply voltage. The fault level will depend on:

— the power available from the source

— the impedance of the network from source to the point of the fault

— the impedance of the fault

— the voltage at the time of the fault

— any stored energy within the HV network.

In addition, the type of fault will have an effect on the calculation as there are several ways in which faults can occur. These include:

— phase to earth

— phase to phase

— phase to phase to earth

— phase to phase to phase.

These represent a large number of variables and so it is normal to assume the worst case; i.e. that the fault has negligible impedance and is a three-phase symmetrical short circuit (phase to phase to phase).

The calculation of a three-phase short circuit current is given by the formula:

$$I_{sc} = \frac{\text{MVA}_s \times 10^3}{V \times 1.732} \tag{5.2}$$

where I_{sc} is the maximum prospective short circuit current (kA), MVA_s is the fault level at the point of supply (MV·A) and V is the voltage (kV).

The fault level at the point of supply can be obtained from the supply authority although, for public network HV supplies, a maximum value of 250 MV·A is normally advised by RECs.

Having assessed the fault level at the intake to the network being considered, the fault level at different points around the network can be calculated. To calculate the fault level accurately, the vector sum of the impedances between the supply and the fault have to be known but this cannot always be found because items such as transformers give percentage impedance.

Therefore it is normal to use percentage impedance when calculating fault levels. To do this all figures have to be converted to a common kV·A rating, known as a base kV·A value (KVA_b). This can be any figure but 1000 kV·A or the relevant transformer size is usually selected for ease of calculation.

The formulae for calculating the percentage impedance for the various items of plant are as follows.

For the source:

$$\text{Source impedance reactance (\%)} = \frac{\text{KVA}_b \times 10^2}{\text{MVA}_s \times 10^3} \tag{5.3}$$

where KVA_b is the base kV·A value (kV·A) and MVA_s the fault level given by the supply authority (MV·A).

For cables:

$$\text{Cable impedance reactance (\%)} = \frac{\text{KVA}_b\,(X\,l)\times 10^5}{V^2}$$

$$(5.4)$$

where KVA_b is the base kV·A value (kV·A), X is the impedance per unit length of the cable (Ω·m^{-1}), l is the length of the cable (m) and V is the phase-to-phase voltage (V).

Equation 5.4 applies for both the HV and LV cables.

For a transformer:

$$\text{Transformer impedance reactance (\%)} = \frac{\text{KVA}_b\times Z}{\text{TR}}$$

$$(5.5)$$

where Z is the normal reactance of the transformer (see section 6.4.8) (%) and TR is the transformer rating (kV·A).

The fault level is given by:

$$\text{MVA}_f = \frac{\text{KVA}_b\times 10^2}{\text{Total \% impedance}\times 10^3}$$

$$(5.6)$$

where MVA_f is the fault level (MV·A) and the total percentage impedance reactance is the sum of the source, transformer and cable percentage impedance reactances (%).

From this, the short circuit current (kA) at the fault will be:

$$I_{sc} = \frac{\text{MVA}_f\times 10^3}{V\times 1.732}$$

$$(5.7)$$

Example: fault level calculations

Consider the simple HV/LV network shown in Figure 5.8

To simplify the calculation, select the base level to be the same as the transformer. Therefore:

$$\text{Source impedance} = \frac{800\times 10^2}{250\times 10^3} = 0.32\%$$

From manufacturer's cable data, 50 mm^2 XLPE/SWA/PVC cable has an impedance of 0.505×10^{-3} Ω·m^{-1}. Therefore, a 15-metre length of cable will have a percentage impedance of:

$$\text{HV cable imp.} = \frac{800\times 0.505\times 15\times 10^{-3}\times 10^5}{11\,000^2}$$

$$= 0.005\%$$

$$\text{Transformer imp.} = \frac{800\times 4.75}{800} = 4.75\%$$

Using two single core 400 mm^2 copper cables in parallel per phase, from manufacturer's cable data, each 400 mm^2

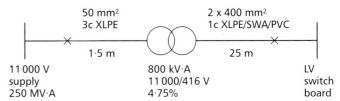

Figure 5.8 A typical HV/LV radial circuit

XLPE/SWA/PVC cable has an impedance of 0.110×10^{-3} ohms/m. A 25 metre length of paralleled cable will have a percentage impedance of:

$$\text{LV cable imp.} = \frac{1}{2}\times\frac{800\times 0.11\times 10^{-3}\times 25\times 10^5}{416^2}$$

$$= 0.636\%$$

Total impedance is:

$$\text{Total imp.} = 0.32 + 0.005 + 4.75 + 0.636 = 5.71\%$$

From equation 5.6, the fault level at the LV switchboard is:

$$\text{MVA}_f = \frac{800\times 10^2}{5.71\times 10^3} = 14.01\ \text{MV·A}$$

Hence, the maximum short circuit current is:

$$I_{sc} = \frac{14.01\times 10^3}{400\times 1.732} = 20.22\ \text{kA}$$

It can be seen from the calculation that the transformer impedance has the greatest effect on the prospective fault level and the size and length of cables have the least effect.

On the basis of the calculation, the designer must ensure that the LV switchgear is cable of withstanding a short circuit current in excess of 20.22 kA for the duration of the fault to prevent any damage or danger arising from such a fault. In addition, the switchgear must incorporate protection devices in the HV and LV switchgear to safely isolate any fault that occurs in the shortest possible time.

5.6.3   Discrimination

Discrimination is the ability of protective devices to detect and correctly respond to a fault condition, so that the faulty circuit or apparatus is isolated with the minimal disruption to any other electrical circuits or apparatus.

Two methods used to obtain discrimination are:

—   *Time grading*: where the protective devices are selected and set to operate faster the further they are from the source of supply. This is also known as unrestricted protection.

—   *Unit protection*: where the protection is only given to the apparatus with which the protective device is associated. This is also known as restricted protection.

Discrimination needs to be designed to ensure that protective devices operate under all condition of overcurrent (i.e. phase-to-phase) faults and phase-to-earth faults.

5.6.3.1   Time graded protection systems

The types of HV devices used in time graded protection systems are:

— high rupturing capacity (HRC) fuses

— AC trip coils fitted with time limit fuses (TLFs)

— inverse definite minimum time (IDMT) relays.

The types of LV devices used in time graded protection systems are:

— high rupturing capacity (HRC) fuses

— inverse definite minimum time (IDMT) relays

— moulded case circuit breakers.

Originally, IDMT relays were of the electromechanical type using a rotating disc to operate contacts. The relays were available in three time–current characteristics (standard, very inverse and extremely inverse), the curves for which were defined by the relevant British Standard in force at the time. In the event that a different characteristic was required, it was necessary to change the relay for one of a different characteristic.

Static protection relays began to be introduced in the 1970s. These perform the same function as the electro-mechanical relays but operate with a greater degree of accuracy and allow characteristic changes to be made without the need to replace the relay.

The time–current characteristic for each protective device in series in a circuit is entered on a discrimination grading chart such as that shown in Figure 5.9, and the relay settings adjusted as necessary to ensure that discrimination is achieved. Modern circuit protection software is available that automatically generates these character-istics.

However, where a designer chooses to carry out the assessment manually, it is normal to use the following time separation in obtaining the necessary time intervals between protective devices:

— a relay in series with another relay should have a time differential of 0.4 seconds

— a relay in series with the new cartridge type of TLF should have a time differential of 0.4 seconds

— a relay in series with an HRC fuse should have a time differential of 0.2 seconds

— a TLF in series with an HRC fuse should have a time differential of 0.2 seconds

— an HRC fuse in series with an HRC fuse should achieve discrimination by comparing the 'let-through' energy of each HRC fuse, which is available from the manufacturers. The total energy required to operate the smaller fuse should be less than the pre-arcing energy of the larger fuse

— a TLF in series with another TLF is not recom-mended because there are problems in obtaining the necessary discrimination at high values of fault level.

The time separations given above should be obtained over the complete operating curve up to the likely fault level.

5.6.3.2   Discrimination across HV/LV transformers

An LV fuse or IDMT/MCCB characteristic can be plotted on an HV discrimination grading chart by making allowances for the differences in voltage (i.e. divide the LV current by 11 000/416). This method is acceptable for a three-phase fault on the LV side but problems can occur because a two-phase fault on the LV side will be approximately 0.866 of the equivalent current on the HV side.

Therefore when checking the discrimination between the HV and LV for a two phase fault it is normal to:

— Plot the HV fuse characteristics on the time–current grading chart.

— Draw the LV device characteristics plus 10% multiplied by 0.866, and check that the LV device is below the HV fuse at the fault level. (10% is added because the characteristics are the mean values for the devices, but the devices can operate within ±10% of these mean characteristics.)

When designing an installation it is possible to calculate the maximum fuse/relay that it is possible to install on the LV side of the distribution.

The maximum LV device can be selected by ensuring that the LV device operating time is less that the HV fuse or relay operating time, i.e:

$$\text{LV device operation} < \frac{\text{HV fuse current}}{0.05} \qquad (5.9)$$

The figure 0.05 has been selected because it is approximately equal to the fuse tolerance (1.1) times the LV fault factor (1/0.866) times the transformation ratio (416/11000).

5.6.4   High rupturing capacity (HRC) fuses

High voltage HRC fuses are normally fitted in switch fuses or fuse switches for the protection of transformers up to the maximum rating of:

— 11 kV, 1000 kV·A

— 6.6 kV, 500 kV·A

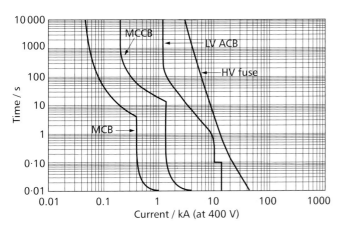

Figure 5.9 Protection grading chart

— 3.3 kV, 300 kV·A

Generally an HRC fuse is only suitable for fault protection because of the difference between its current rating and its minimum fusing current. Typically, the minimum fusing current for an HRC fuse is 1¹/₂ to 2 times the rated current for the fuse.

If a fuse is required to carry for any long period a current greater than its rated current, but less than its minimum fusing current, deterioration may occur which could result in unwanted or incorrect operation at a later time.

In selecting a suitable fuse, consideration should be given to the following factors:

—   maximum load current

—   transformer magnetising in-rush current (the actual values of in-rush currents are impossible to predict because of the instantaneous value of the voltage at the time of switch-on; a general figure used in designing a network is that the fuse should withstand an in-rush of 10 times the full load current for 0.1 seconds)

—   the fuse should protect the installation, consisting of the HV cable, HV/LV transformer and the LV cabling.

When specifying fuses it is important to state the medium in which the fuses will be required to operate, i.e. air or oil.

When looking to replace or install a fuse for the first time, the following items need to be considered:

—   the current rating

—   the system voltage

—   the system fault level

—   the time–current characteristics of the fuse

—   the $I^2 t$ characteristics of the fuse

—   if a striker pin is required

—   the physical dimensions of the fuse

—   the maximum size of fuse that the switch fuse can hold

—   if replacing a ruptured fuse then all three phase fuses should be replaced.

5.6.5   Trip coils and time limit fuses (TLF)

Over-current trip coils are normally fitted on circuit breakers which supply:

—   radial feeders

—   ring feeders in open ring configurations

—   transformer feeders up to 1000 kV·A.

The basic connection of an over-current trip coil and time limit fuse is shown in Figure 5.10

The device consists of the TLF in parallel with a trip solenoid which has a central plunger. The plunger is arranged to operate the circuit breaker tripping

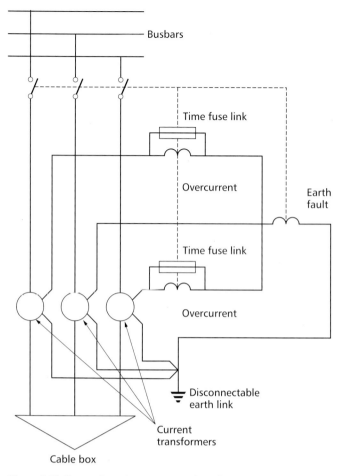

Figure 5.10 Connection of over-current trip coil and time limit fuse

mechanism. The solenoid or over-current trip coil is instantaneous in operation for all currents in excess of the coil rating. The device is given an inverse time characteristic by having the TLF in parallel. The fuse connected in parallel has a low impedance and so carries most of the current from the current transformer (CT) and prevents the coil from operating until the fuse ruptures.

The manufacturers of time limit fuses publish time–current characteristics which can be multiplied by the CT ratio to give the time–current characteristics for the circuit breaker fitted with the TLF fuse.

5.6.6   Inverse definite minimum time (IDMT) relays

IDMT relays are normally fitted to circuit breakers which supply:

—   distribution transformers

—   incoming supply

—   generator supply

—   ring feeders in open ring configuration.

The IDMT relays can provide just over-current or over-current and earth fault protection, just like the TLFs and trip coil devices. However greater adjustment is available of the operating characteristic.

Other facilities on the IDMT relay are:

—   *instantaneous settings*: where the relay operates instantaneously on any current above the setting

— *definite time setting*: where the relay will operate for a definite time for any current in excess of the instantaneous current setting.

With all the terms now defined a standard table can be used when carrying out the grading of IDMT relays.

When determining the time–current characteristics for the IDMT relay the following information is required:

— the circuit breaker rating

— the CT ratio.

From this information, the maximum current I_n is defined and, from this figure, switches are set on the relay to select the required protection curve to provide discrimination with downstream protective devices.

5.6.7 Unit protection

Unit protection works on the principle that if the current entering and leaving the apparatus is the same then the apparatus is considered healthy. If there is a difference in the two current readings, the apparatus is considered to have a fault. As unit protection systems are stable they have low trip settings and fast operating times. Some types of unit protection are:

— restricted earth fault relays

— balanced voltage relay

— circulating current relay.

Unit protection is a relatively specialist subject and is unlikely to be encountered by the building services engineer.

5.6.7.1 Earth fault protection

Earth fault protection is designed to operate for current levels less than full load, when there is a fault between any phase and earth.

Where the CTs of the earth fault relays are connected determines the zone of protection that the earth fault relay provides. The two types of earth fault protection are:

— *restricted earth fault protection*: the protected zone is the windings of either a transformer or an alternator

— *unrestricted earth fault protection*: the zone of protection is all the external wiring down stream of the CT connection.

5.6.7.2 Balance voltage relays

These relays are designed to protect a particular cable section on a distribution system. The relay only operates if there is a fault in the cable section that the relay is protecting.

Considering the closed ring arrangement in Figure 5.4, under normal conditions or for through faults the voltage generated by the relay CTs at each end of the cable will be equal but opposite in direction. Therefore no current will flow through the pilot wires and so the trip coils will not operate.

If there is an internal fault in the cable section, the current flowing through each circuit breaker will be different. The voltage generated by the two CTs will be unbalanced and current will flow through the pilot wires causing the trip coils to operate and open the supply circuit breakers.

The systems which use this principle are Solkor A, Solkor B and Translay.

5.6.7.3 Circulating current relays

Again, these relays are designed to protect a particular cable section on a distribution system. The relay only operates if there is a fault in the cable section that the relay is protecting.

The trip coils are connected in parallel to the CTs at a point that is at the same potential from each CT.

Under normal conditions or for through faults, the current generated in each CT is equal and, as the trip relay is connected to the electrical midpoint, no current flows. Therefore the trip relay does not operate.

If there is an internal fault in the cable section the current flowing through each circuit breaker will be different. Therefore the current generated by the two CTs will be unbalanced and so the trip coil will no longer be at the electrical midpoint. Therefore the trip coil will operate and so disconnect the supply circuit breakers. This is illustrated in Figure 5.11.

A system which uses the circulating current principle is the Solkor R relay.

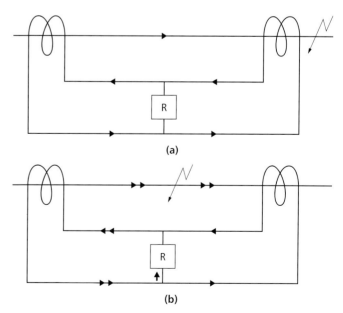

Figure 5.11 Basic arrangements for unit protection; (a) current distribution with 'through fault', (b) current distribution with 'in-zone fault' resulting in unequal currents from current transformers

5.6.8 Intertripping of HV and LV circuit breakers

It is recommended that, on supplies over 1 MV·A, a system of intertripping be incorporated into the electrical design. Depending on the type of supply it may be beneficial to provide this arrangement on smaller sized supplies. The idea behind intertripping is to ensure that the HV breaker does not close onto a live system.

This requirement is normally carried out with auxiliary contacts on the circuit breaker, relays and 'send' and 'receive' signalling cables. Using this arrangement, when the HV circuit breaker trips (under fault etc.) a signal is sent which also trips the LV breaker on the other side of the transformer. Similarly, if the LV breaker trips (under fault etc.) then a signal is sent to trip the HV breaker.

5.7 HV cables

5.7.1 Cable construction

HV cables are produced in a wide variety of types to suit various applications. The construction of a typical HV cable is shown in Figure 5.12, which illustrates the various layers as follows:

(a) A conductor core of stranded copper or solid aluminium in standard sizes of cross-sectional area (CSA). Manufacturers' cable data should be consulted for the current carrying capacities of the different cable sizes.

(b) Insulation of either cross-linked polyethylene, (XLPE), cross-linked ethylene propylene rubber (EPR), polyvinyl chloride (PVC) or paper impregnated with insulating compound. XLPE is usually specified for normal industrial applications as this has a higher operating temperature than PVC. However, EPR may be required where the cable is continuously subject to wet conditions.

(c) Semiconducting insulating screen.

(d) Metallic screen of overlapped copper tape to provide an earth fault current path. In some types of cable, this layer comprises a lead sheath

(e) Filler to form a circular section.

(f) Bedding sheath of PVC.

(g) Armouring of galvanised steel wires or tape to provide mechanical protection and an earth fault current path.

(h) An oversheath of PVC, reduced flame propagation PVC (RPPVC), low smoke zero halogen compound (LS0H) or medium density polyethylene (MDPE). MDPE is specified where greater permeability to moisture and greater resistance to abrasion is required but this type of sheath is not recommended in areas of fire risk.

Thus a typical cable may be specified as 3×70 mm^2 Cu, XLPE, CTS, PVC, SWA, LS0H, 11 kV. This cable would have three copper conductors of 70 mm^2 CSA with XLPE insulation, a copper tape screen, PVC bedding, steel wire

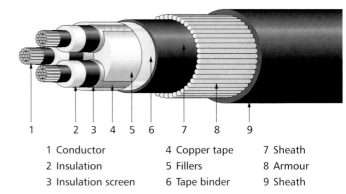

1 Conductor	4 Copper tape	7 Sheath
2 Insulation	5 Fillers	8 Armour
3 Insulation screen	6 Tape binder	9 Sheath

Figure 5.12 Construction of a typical HV cable

armouring and a low smoke, zero halogen outer sheath. The cable would be rated for a phase-to-phase voltage of 11 kV.

Cables are often identified by a four-letter code that identifies the number of conductors, the conductor type, the type of screen and whether the cable is armoured.

Table 5.1 gives details of different cables that can be used for HV installations. This table is not exhaustive and cable manufacturers should be consulted for advice on cables for specific applications.

5.7.2 Selection of cables

The cables should be selected by using manufacturers' design information and considering the following factors:

— operating voltage

— normal full load current

— prospective short circuit fault rating,

— derating factors due to:

   (a) thermal resistivity of soil

   (b) ground temperature

   (c) ambient air temperature

   (d) depth that the cable is to be installed

   (e) grouping and proximity of other cables

— prospective fault current and the type of system protection

— voltage drop

— degree of mechanical protection required and the method of installation

— overall diameter, flexibility and minimum bending radius

— type of termination.

5.7.3 Testing of cables

All HV cables need to be pressure tested prior to commissioning and after repairs or alterations. Such testing must be undertaken by an HV 'authorised person', see section 5.8.1.

The pressure test should only be carried out after an insulation test (1000 V) has been carried out between each

Table 5.1 Types of cable used in HV installations

Designation	Cable description	Relevant standard
—	Copper or aluminium conductor, paper insulated, screened, lead alloy sheathed, single wire armoured with/without serving, PVC oversheath	BS 6480[20]
—	Copper or aluminium conductor, paper insulated, belted, lead alloy sheathed, single wire armoured with/without serving, PVC oversheath	BS 6480[20]
SCTA	Single core, copper conductor, XLPE/EPR insulated, copper tape screened, single wire armoured, PVC/RPPVC/LS0H/MDPE oversheath	BS 6622[22]
TCTA	Three-core, copper conductor, XLPE/EPR insulated, copper tape screened, single wire armoured, PVC/RPPVC/LS0H/MDPE oversheath	BS 6622[22]
SATA	Single core, aluminium conductor, XLPE/EPR insulated, copper tape screened, single wire armoured, PVC/RPPVC/LS0H/MDPE oversheath	BS 6622[22]
TATA	Three-core, aluminium conductor, XLPE/EPR insulated, copper tape screened, single wire armoured, PVC/RPPVC/LS0H/MDPE oversheath	BS 6622[22]
SCTU	Single core, copper conductor, XLPE/EPR insulated, copper tape screened, unarmoured, PVC/RPPVC/LS0H/MDPE oversheath	IEC 502[33]
TCTU	Three-core, copper conductor, XLPE/EPR insulated, copper tape screened, unarmoured, PVC/RPPVC/LS0H/MDPE oversheath	IEC 502[33]
SATU	Single core, aluminium conductor, XLPE/EPR insulated, copper tape screened, unarmoured, PVC/RPPVC/LS0H/MDPE oversheath	IEC 502[33]
TATU	Three-core, aluminium conductor, XLPE/EPR insulated, copper tape screened, unarmoured, PVC/RPPVC/LS0H/MDPE oversheath	IEC 502[33]
SCWU	Single core, copper conductor, XLPE/EPR insulated, copper wire screened, unarmoured, PVC/RPPVC/LS0H/MDPE oversheath	BS 6622[22]
TCWU	Three-core, copper conductor, XLPE/EPR insulated, copper wire screened, unarmoured, PVC/RPPVC/LS0H/MDPE oversheath	BS 6622[22]

conductor, and each conductor and earth. It is normal practice for this test to be carried out for a duration of not less than one minute.

The HV pressure test will use a DC voltage of the value specified in BS 6480[20] and applies to paper insulated cables. The voltage must be applied and increased gradually to the full value and maintained at that value for 15 minutes.

When an HV pressure test has to be carried out on a mixture of old and new cables then the test voltage for the old cables should be used.

When testing cables the age of switchgear may limit the maximum permissible test voltage. The test voltage for the cable should be compared with the maker's recommendations for the switchgear and the lower test voltage of the two applied.

Tests on cable should normally always be carried out with a double isolation. If it is impracticable to achieve this then the following should be followed:

— The test voltage must not exceed the value specified for old cables.

— All access to live equipment must be locked off and danger notices displayed.

— Should a breakdown occur then no further tests should be carried out until the cable or equipment has been completely isolated with a double break.

— If double isolation is available but at only one end, then the test should be applied at the end with double isolation.

When the test is finally completed, care should be taken to ensure that the cable has discharged prior to handling the test connections. If tests are carried out via switchgear with earthing facilities, the cable should be discharged by connecting it to earth.

Further information on testing is contained in section 12.

5.8 Safety aspects

The Electricity At Work Regulations 1989[29] are designed to require precautions to be taken against the risk of death or personal injury from electricity in work activities.

Regulation 4, which is split into four parts, covers the aspects of electrical systems and equipment which are fundamental to safety.

When working on HV it is necessary to ensure that personnel have the necessary experience and training for the tasks that they are required to perform. It is normal to distinguish such staff as authorised and competent persons.

5.8.1 Authorised person (AP)

An authorised person should have the following qualifications and qualities:

— be technically competent

— be familiar with the electrical distribution system for which he/she is authorised

— have adequate training on the system and work which he is required to perform

— been appointed by a suitably qualified HV engineer

— be able to recognise danger

— be familiar with first aid, including resuscitation

— be familiar with the site/organisation, electrical safety rules and relevant electrical legislation

— be methodical and level headed.

5.8.2 Competent person (CP)

A competent person should have the following qualifications:

— have sufficient technical knowledge and experience to avoid danger

— be competent for the duties to be performed

— been appointed by a suitably qualified HV engineer

— be familiar with first aid, including resuscitation.

5.8.3 Access and control of work

All substations on a site should be under the control of an authorised person (AP) and all work to be carried out in the substations must be addressed to the AP.

The AP should be aware at all time of any work that is being carried out in a substation under his/her control. The AP should control the access to the substation and should have the power to refuse entry to the substation to any person, regardless of position or status, who does not have a valid reason for entry.

When a substation is unattended it should be kept locked. No person other than an authorised person, or a competent person working under the direct control of an AP, should carry out any work on any HV equipment; and then only when the AP in charge of the substation where the work is to be carried out has been notified and given his/her permission.

When contractors are employed to carry out work on HV equipment, the contractor must appoint authorised and competent persons to carry out the work. At all times these persons must ensure that the safety regulations are complied with.

A new installation should be considered to be a substation on reaching the stage of completion where any part of the equipment is, or could be energised by normal operation of the equipment and this results in a potential exceeding 650 V (phase-to-earth) being developed within the equipment.

No person should carry out work on HV equipment alone.

A log book should be provided to record the following:

— all switching operations

— the issue and cancellation of all permits to work

— the handover of system responsibility from one authorised person to another

— the receipt of any operational restriction notices (e.g. notices that highlight/advise of possible defects with equipment).

5.8.4 Permit to work

A permit to work should be issued whenever work is being carried out on HV equipment. Before the work is carried out, the equipment should be:

— isolated from the supply

— securely locked off so that it cannot be energised

— tested to ensure that the equipment is dead

— earthed

— have 'CAUTION — DO NOT INTERFERE' warning notices placed where the equipment is isolated

— have 'DANGER — LIVE APPARATUS' warning notices placed on live adjacent equipment.

The permit to work should clearly indicate the following:

— the date and time that it was issued

— the competent person that it was issued to

— the authorised person in charge of the substation who issued the permit to work

— the address, location and identity of the equipment to be worked on

— the location and identity of where the equipment has been isolated

— the location of where the equipment is earthed, and whether the earth can be removed for testing

— the locations where 'DANGER' or 'CAUTION' notices are displayed

— the work or test which is to be carried out on the piece of equipment

— the signatures of both the issuing AP and receiving CP stating that they understand the work that is being carried out and that they confirm that the equipment is safe to work on

— an area where the permit can be signed-off, giving details that the work has been completed

— details of any other permits in existence in the vicinity of the work being undertaken.

5.8.5 Safety equipment

Safety equipment should be provided and regularly tested in accordance with the manufacturer's instructions, and available at all times to the persons requiring it.

Safety equipment should always be worn and used whenever necessary to avoid danger. Safety equipment should include, rubber gloves, rubber mats and isolation padlocks.

Safety equipment should be inspected by the user for visible defects every time that it is used. Any suspect item should be withdrawn and replaced.

Safety equipment should be stored in accordance with the manufacturer's recommendations.

5.8.6 Earthing equipment

Only the earthing equipment provided by the equipment manufacturer should be used. The equipment should be tested before use.

5.8.7 Testing equipment

A suitable HV indicator, voltage tester or other equipment should be used to ensure that equipment is dead. The test equipment should be tested by using a proving unit before and after use.

5.8.8 Safety rules to be used when working on HV

The following rules must be observed when working on HV installations:

(1) People are responsible for their own safety and the safety of those around them.

(2) There is always time for safety. Never take risks to save time.

(3) Never work without a valid permit to work.

(4) Never rely on the word of somebody else, see for yourself.

(5) Never work alone.

(6) Refuse to carry out work that you consider dangerous, that you do not understand or that you have not been trained for.

(7) Danger signs and barriers are put there as a warning; do not interfere with them.

(8) Never leave an unlocked substation unattended.

(9) Make sure that you fully understand the work to be performed.

(10) Ensure that rubber mat are in position in all places where work is to be performed.

(11) Remove any metal or loose objects from pockets before work commences.

(12) Always remember the basic principles when working on HV equipment:

— MAKE SURE IT IS DEAD: by disconnecting the equipment to be worked on from the load and from the supplies.

— MAKE SURE IT IS DEAD: by testing at all points that have been or could have been live. Use an HV indicator to prove that the equipment is dead. Afterward confirm your findings with an earthing stick.

— MAKE SURE THAT IT STAYS DEAD: by earthing down at the point of work.

(13) Ensure that every operation, test or check is observed by another AP or CP.

(14) Never work on any HV equipment after the permit to work has been cancelled.

5.9 Operation and maintenance

5.9.1 General

The Electricity at Work Regulations 1989[29] cover maintenance under Regulation 4(2). These require that:

(a) maintenance is carried out in order to ensure the safety of the system

(b) the quality and frequency of maintenance is sufficient to prevent danger

(c) records of maintenance is kept through out the working life of an electrical system

(d) regular inspection of plant is carried out.

Planned maintenance is essential to ensure continuity of service and economy of labour and material. BS 3811: *Maintenance terms in terotechnology*[30] gives a combination of management financial, engineering and other practices applied to physical assets in pursuit of economic life cycle costs.

The complete planned maintenance and operation system should embrace the following:

— a detailed list of local instruction, which itemises how to perform switching operations and testing operations to make a piece of equipment dead and safe to work on

— a job sheet, which provides information on the work to be done

— a planning chart, which provides overall planning of the work

— a permit to work, which issues orders for the work to be done.

The frequency and details of maintenance requirements will depend on the manufacturers' recommendations. Additional information on the inspections, operational checks and examination of HV switchgear can be obtained from BS 6626[23].

5.9.2 Maintenance period

The safety of people and buildings depends on the correct operation of equipment such as protection relays, circuit breakers and automatic fire extinguishing systems. This essential equipment must be regularly maintained, and details of the work recorded.

The following routine intervals may help as a general guide for maximum values and they can be shortened at the discretion of a suitably qualified HV engineer following consideration of such items as:

— the environmental conditions

— the degree of normal use

— the number of operations under fault conditions

— historical information on the most appropriate maintenance period.

Every 13 weeks:

— Check HV accommodation.

— Check safety equipment and tools.

— Inspect LV control battery unit.

— Check and record instrument readings.

— Visually inspect all equipment externally for signs of distress/deterioration.

Every 52 weeks:

— Check fire suppression system (if installed).

— Inspect and operate circuit breakers or fuse switches.

— Inspect and operate LV switchgear associated with transformer output.

— Inspect transformers and HV cubicles.

— Clean out any air cooled transformers.

Every 156 weeks:

— Test HV earth electrode system.

— Test transformer oil.

— Check protection equipment

Every 312 weeks:

— Pressure test automatic fire extinguishing system.

— Examine oil circuit breakers and change oil.

— Examine vacuum circuit breakers.

— Examine SF_6 circuit breakers.

— Examine oil filled switches and change oil.

When working normally on HV equipment, ensure that:

— only suitably qualified personnel operate the equipment

— all safety warning notices are complied with, and that all enclosures are closed and all covers in place

— all operatives are trained to recognise indications of mal-operation or malfunction and are aware of the actions to take in the event of such indications.

When personnel are carrying out maintenance operations they should:

— comply with all safety working procedures

— be fully conversant with all information on measures relating to personnel safety

— ensure that work is only carried out on the equipment for which a permit to work has been issued.

References

1   BS EN 50187: 1997: *Gas-filled compartments for a.c. switchgear and controlgear for rated voltages above 1 kV and up to and including 52 kV* (London: British Standards Institution) (1997)

2   BS EN 60129: 1994 (IEC 60129: 1984): *Specification for alternating current disconnectors and earthing switches* (London: British Standards Institution) (1994)

3   BS EN 60255: *Electrical relays* (16 parts) (London: British Standards Institution) (1995–2003)

4   BS EN 60265: *Specification for high-voltage switches*; Part 1: 1998 (IEC 60265-1: 1998): *Switches for rated voltages above 1 kV and less than 52 kV* (London: British Standards Institution) (1998)

5   BS EN 60282: *High-voltage fuses*: Part 1: 2002: *Current-limiting fuses* (London: British Standards Institution) (2002)

6   BS EN 60298: 1996 (IEC 60298: 1990): AC *metal-enclosed switchgear and controlgear for rated voltages above 1 kV andd up to and including 52 kV* (London: British Standards Institution) (1996)

7   BS EN 60420: 1993 (IEC 60420: 1990): *Specification for high-voltage alternating current switch-fuse combinations* (London: British Standards Institution) (1993)

8   BS EN 60470: 2001 (IEC 60470: 2000): *High-voltage alternating current contactors and contactor-based motor starters* (London: British Standards Institution) (1990)

9   BS EN 60644: 1993 (IEC 60644: 1979): *Specification for high voltage fuse-links for motor circuit applications* (London: British Standards Institution) (1993)

10 BS EN 60694: 1997 (IEC 60694: 1996): *Common specifications for high-voltage switchgear and controlgear standards* (London: British Standards Institution) (1997)

11 BS EN 61330: 1996 (IEC 61330: 1995): *High-voltage/low-voltage prefabricated substations* (London: British Standards Institution) (1994)

12 BS EN 62271: *High-voltage switchgear and controlgear* (4 parts) (London: British Standards Institution) (2001–2004)

13 BS 159: 1992: *Specification for high-voltage busbars and busbar connections* (London: British Standards Institution) (1992)

14 BS 923: *Guide on high-voltage testing techniques*: Part 1: 1990 (IEC 60060-1: 1989): *General* (London: British Standards Institution) (1990)

15 BS EN 60044-1: 1999 (IEC 60044-1: 1996): *Instrument transformers. Current transformers* (London: British Standards Institution) (1990)

16 BS 3941: 1975: *Specification for voltage transformers* (London: British Standards Institution) (1975)

17 BS 5207: 1975: *Specification for sulphur hexafluoride for electrical equipment* (London: British Standards Institution) (1975)

18 BS 5311: 1996: *High-voltage alternating-current circuit-breakers* (London: British Standards Institution) (1996)

19 BS 5992: *Electrical relays*: Part 1: 1980 (IEC 60255-0-20: 1974): *Specification for contact performance of electrical relays*; Part 4: 1983 (IEC 60255-14:1981): *Specification for contact loads of preferred values used in endurance tests for electrical relay contacts*; Part 5: 1983 (IEC 60255-15: 1981): *Specification for test equipment used in endurance tests for electrical relay contacts*; Part 6: 1984: *Specification for the basic modules for the dimensions of general purpose all-or-nothing relays* (London: British Standards Institution) (1980–1984)

20 BS 6480: 1988: *Specification for impregnated paper-insulated lead or lead alloy sheathed electric cables of rated voltages up to and including 33 000 V* (London: British Standards Institution) (1988)

21 BS 6553: 1984 (IEC 60787:1983): *Guide for selection of fuse links of high-voltage fuses for transformer circuit applications* (London: British Standards Institution) (1984)

22 BS 6622: 1999: *Specification for cables with extruded cross-linked polyethylene or ethylene propylene rubber insulation for rated voltages from 3.8/6.6 kV up to 19/33 kV* (London: British Standards Institution) (1999)

23 BS 6626: 1985: *Code of practice for maintenance of electrical switchgear and controlgear for voltages above 1 kV and up to and including 36 kV* (London: British Standards Institution) (1985)

24 BS 6878: 1988 (EN 50052: 1986): *Specification for high-voltage switchgear and controlgear for industrial use. Cast aluminium alloy enclosures for gas-filled high-voltage switchgear and controlgear* (London: British Standards Institution) (1988)

25 BS 7197: 1990: *Specification for performance of bonds for electric power cable terminations and joints for system voltages up to 36 kV* (London: British Standards Institution) (1990)

26 BS 7315: 1990 (EN 50064: 1989): *Specification for wrought aluminium and aluminium alloy enclosures for gas-filled high-voltage switchgear and controlgear* (London: British Standards Institution) (1990)

27 BS 7835: 2000: *Specification for armoured cables with extruded cross-linked polyethylene or ethylene propylene rubber insulation for rated voltages from 3.8/6.6 kV up to 19/33 kV having low emission of smoke and corrosive gases when affected by fire* (London: British Standards Institution) (2000)

28 The Electricity Safety, Quality and Continuity Regulations 2002 Statutory Instruments 2002 No. 2665 (London: The Stationery Office) (2002)

29 The Electricity at Work Regulations 1989 Statutory Instruments 1989 No. 635 (London: The Stationery Office) (1989)

30 BS 3811: 1993: *Glossary of terms used in terotechnology* (London: British Standards Institution) (1993)

31 IEC 60502: *Power cables with extruded insulation and their accessories for rated voltages from 1 kV (Um = 1,2 kV) up to 30 kV (Um = 36 kV)*: Part 1: 2004: *Cables for rated voltages of 1 kV (Um = 1,2 kV) and 3 kV (Um = 3,6 kV)*; Part 2: 1998: *Cables for rated voltages from 6 kV (Um = 7,2 kV) up to 30 kV (Um = 36 kV)*; Part 4: 1997: *Test requirements on accessories for cables with rated voltages from 6 kV (Um = 7,2 kV) up to 30 kV (Um = 36 kV)* (Geneva: International Electrotechnical Commission) (1997–2004)

6 Transformers

6.1 Introduction

The purpose of a transformer is to convert, i.e. transform, electrical energy from one voltage to a higher or lower voltage with a corresponding decrease or increase in current respectively.

Transformers are used in many applications from small domestic products like televisions and hi-fi systems at very low voltages to electrical transmission distribution systems via overhead power lines at many hundreds of thousands of volts.

It is not the intention of this document to cover all possible applications for transformers, rather those that are likely to be encountered by the building services engineer in the course of a typical, large project. Power transmission transformers (i.e. those transforming voltages between 400 kV and 11 kV) are specialist in their nature and will normally be specified and installed by National Grid or a regional electricity company (REC). This section will, therefore, concentrate on power transformers that are installed to convert a supply voltage of 11 kV received from a REC to low voltage, or to convert one low voltage to another.

Transformers used for instrumentation, lighting and other applications are considered to be outside the scope of this Guide.

6.2 Fundamental principles of transformers

A transformer consists essentially of a magnetic core built up of steel laminations, upon which are wound two sets of coils for each phase, suitably located with respect to each other, and termed the primary and secondary windings respectively, see Figure 6.1.

The number of turns within each coil will determine the voltage ratio between the primary and secondary windings. Where the secondary voltage is lower than the primary voltage, the transformer is termed a 'step-down' transformer; where the secondary voltage is higher than the primary voltage, the transformer is termed a 'step-up' transformer.

The primary winding is that winding that is connected to the supply, irrespective of whether the transformer is of the step-up or step-down type. The secondary winding is always that winding that is connected to the load.

The transformer operates using the principle of electro-magnetic induction. When an alternating voltage is applied to the primary winding, an alternating magnetic flux is generated in the core. This in turn generates a voltage in the secondary winding. This principle applies whether the transformer is single phase (not a normal arrangement in building power systems) or three-phase. The voltage generated is determined by the number of turns in the primary and secondary windings. Increasing the number of turns in the secondary winding increases the voltage produced.

Nearly all power transformers encountered in building power systems are three-phase units. These normally have the primary windings connected as a delta arrangement with each phase of the incoming supply connected to two windings. Since the secondary loads on the transformer are rarely balanced between phases, it is normal for secondary windings to be connected in a star config-uration with each phase of the load connected to a single winding and for the other ends of the three winding to be connected together and brought out as a neutral connection. This is shown diagrammatically in Figure 6.2.

The capacity of a power transformer is defined by the full-load power, in kV·A, that it can produce. For single-phase transformers, this is expressed as:

$$S = U_{\text{ph}} I_{\text{ph}} \times 10^{-3} \tag{6.1}$$

where S is the full load power (kV·A), U_{ph} is the secondary (no-load) voltage (V) and I_{ph} is the rated output current (A).

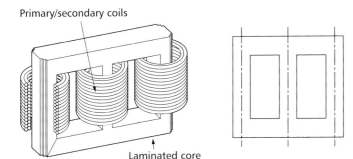

Figure 6.1 Magnetic circuit design for a typical 3-phase transformer

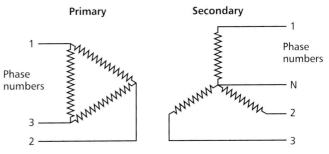

Figure 6.2 Diagrammatic representation of a delta-star transformer

For three-phase transformers, this is expressed as:

$$S = U_{ph} I_{ph} \times 10^{-3} \times \sqrt{3} \qquad (6.2)$$

The values for voltage and current in star- and delta-connected three-phase windings are as follows.

For star connection:

$$\text{Phase current} = \frac{S \times 10^{-3}}{\sqrt{3} \times U_{ph}} \qquad (6.3)$$

$$\text{Phase voltage} = \frac{U_{ph}}{\sqrt{3}} \qquad (6.4)$$

where S is the rated power (kV·A).

For delta connection:

$$\text{Phase current} = \frac{I_{ph}}{\sqrt{3}} = \frac{S \times 10^{3}}{\sqrt{3} \times U_{ph}} \qquad (6.5)$$

$$\text{Phase voltage} = \text{line voltage} = U_{ph} \qquad (6.6)$$

A power transformer is not a perfect transmission device and suffers both 'iron losses' and 'copper losses'. Iron loss is predominantly the energy required to sustain the magnetic field, even when no current is being drawn in the secondary windings. Copper loss is the energy lost due to the winding resistances. The efficiency of power transformers is generally very high with losses typically less than 5%.

When specifying transformers, consideration should be given to the maximum demand in kV·A at which the transformer is expected to operate, making due allowance for future load expansion. It should be noted that transformers are normally rated in kV·A at a power factor of 0.8.

It is important to consider the size of the transformer in relation to the harmonics that will be generated in the loads and that will consequently flow in the secondary windings. These can raise the neutral current to above the rated value and produce a considerable heating effect.

Information on the loads, anticipated load cycle and harmonics should be issued to the transformer manufacturer as part of the specification. Current flowing in the windings creates internal heat and hence expansion of the windings. Where frequent load cycles are likely to occur (e.g. a large motor), the design of the transformer must take account of the stresses arising from frequent motor starts.

6.3   Basis of construction and operation

The design of a typical three-phase power distribution transformer incorporates a laminated iron core of three vertical limbs surmounted by a top and bottom yoke. The conductor windings, which can be either aluminium or copper, are concentrically wound to form a coil around each vertical limb. The whole assembly is held together between top and bottom core clamps. The HV and LV ter-minations are usually brought out on opposite sides of the assembly.

The complete assembly is mounted on a support frame that either forms the roller base for dry-type units or tank spacer for liquid-filled units. Manufacturing techniques vary considerably between liquid-filled units and dry-type transformers but their object or purpose is identical.

6.4   Transformer specification

To enable a manufacturer to design and construct a transformer, it is necessary for the specifier to provide/confirm the following information:

—   design and construction standards

—   type of transformer

—   method of insulation and cooling

—   transformer rating

—   vector group

—   primary and secondary electrical supplies

—   transformer duty

—   location

—   impedance

—   neutral and earthing arrangements

—   tappings

—   terminals and cable boxes

—   accessories and tank fittings

—   enclosures

—   testing requirements.

Expressing transformer details as described above gives simplistic guidelines to provide rating, ratio, vector reference, impedance, protection and cooling. An example of this nomenclature applied to a typical external 1000 kV·A oil-filled transformer would appear as follows:

—   *Rated power*: 1000 kV·A

—   *Ratio*: 11 kV/416 V (primary winding/secondary winding)

—   *Vector group*: Dyn11 (delta/star connected with neutral brought out and 30° phase shift)

—   *Impedance*: 6%

—   *Protection*: IP65

—   *Cooling*: AN (air natural)

—   *Rated insulation level*: 12 kV

—   *Insulation system/class*: O

—   *Haulage/lifting lugs*: Yes

—   *Rollers*: No.

Each of these items will be considered separately in the following paragraphs.

A schedule of information that is required is included as Annex A of BS EN 60076-1[1].

6.4.1 Standards

The selection, construction and use of power transformers is covered by BS EN 60076: *Power transformers*[1], together with a variety of standard specifications, as follows:

— BS EN 60044-1[2]: current transformers

— BS EN 60726[3]: dry-type power transformers

— BS EN 60742/BS 3535-1[4]: isolating transformers

— BS EN 61378[5]: converter transformers

— BS EN 61558[6]: safety of power transformers, power supply units and similar

— BS 171[7]: power transformers

— BS 3941[8]: voltage transformers

— BS 6436[9]: ground mounted distribution transformers

— BS 7452[10]: separating transformers, autotransformers, and variable transformers and reactors

— BS 7628[11]: combined transformers

— BS 7735[12]: loading of oil-immersed power transformers

— BS 7821[13]: three-phase oil-immersed distribution transformers

— BS 7844[14]: three-phase dry-type distribution transformers

— IEC 60354[15]: loading guide for oil-immersed power transformers

— IEC 60905[16]: loading guide for dry-type power transformers.

Prior to the introduction of BS EN 60076[1], the specification for power transformers was BS 171[7], which has largely been superseded.

6.4.2 Types of transformer

There are three main types of transformers that may be encountered in buildings. These are:

(*a*) *Power transformers*: these are defined by BS EN 60076-1[1], clause 3.1.1, as 'a static piece of apparatus with two or more windings which, by electromagnetic induction, transforms a system of alternating voltage and current into another system of voltage and current usually of different values and at the same frequency for the purpose of transmitting power.' The windings of power transformers are in parallel with the associated systems. The transfer of power is solely by induction.

Virtually all power transformers used in buildings are of the delta-star configuration having delta-connected primary windings and star-connected secondary windings, the centre point of which is earthed and connected as the neutral point for the electrical system.

(*b*) *Autotransformers*: these are defined by BS EN 60076-1[1], clause 3.1.2, as 'transformers in which at least two windings have a common part.' The throughput power is transferred partly by conduction and partly by induction.

Autotransformers are available as both star and delta units and provide a solution for either step-up or step-down transformers. Autotransformers can have a cost advantage where the ratio of input to output voltage is less than 5:1.

(*c*) *Booster transformers*: these are defined by BS EN 60076-1[1], clause 3.1.3, as 'a transformer of which one winding is intended to be connected in series with a circuit in order to alter its voltage and/or shift its phase. The other winding is an energising winding.' Their windings are electrically independent; the additional power is transferred purely inductively.

Booster transformers are used in main distribution and transmission systems to maintain a constant supply voltage on the electrical network. This is achieved using motorised tap-changers which vary either the primary or secondary voltage in response to load demand.

The diagrammatic arrangements of these are shown in Figure 6.3.

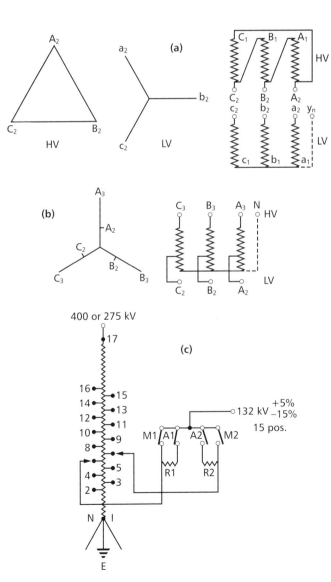

Figure 6.3 Winding configurations; (a) delta-star transformer, (b) autotransformer, (c) booster transformer (1 phase shown)

6.4.3   Insulation and cooling

For most applications in building services, either liquid-filled or dry-type cast resin power transformers are the most commonly used. Such transformers are double wound, i.e. having one primary winding and one secondary winding.

The method of cooling is stated by the manufacturer in the form of four capital letters, the first two letters denoting the coolant and the manner of circulation for the winding, and the last two letters indicating the coolant and manner of circulation for cooling the outside of the transformer. For example:

—   AN: dry-type transformer with natural air circulation

—   ONAN: oil-immersed air-cooled transformer.

Definitions of the code letters are given in Table 6.1.

In the past, polychlorinated biphenyls (PCBs) were used as an insulation medium in transformers and power factor capacitors. PCBs are a class of chlorinated hydrocarbons that have been used extensively since 1930 for a variety of industrial uses.

The value of PCBs derived from their chemical inertness, resistance to heat, non-flammability, low vapour pressure and high dielectric constant. However, PCBs have since been found to exhibit chronic toxicity and are suspected of being carcinogenic such that their use has been discontinued. PCBs have been supplied in commercial formulations under a variety of trade names, including Askarel, Pyrochlor, Ducanol

As a replacement for PCBs, silicone liquid was introduced in the 1970s as a flame-retarding and non-polluting alternative. Other synthetic liquids, such as Midel, with a fire point >300 °C may be encountered besides silicone liquid.

Cooling is absolutely necessary in any transformer as the losses from the core and windings are distributed as heat. Two types of cooling are offered:

—   AN: air through natural convection

—   AF: air forced via mechanical ventilators.

All windings within liquid-filled units are cooled via the insulating medium to radiators, while the windings in dry-type cast resin units are directly cooled to air. The potential to increase the given rating of a particular transformer (possibly for high, short-duration loads) is 15% uplift with liquid-filled units and 40% uplift with dry-type cast resin units. However, it is important for the specifier to agree the circumstances for such uplifts with the client/user during the design process, and to discuss how this is to be provided with the transformer supplier (usually by forced ventilation).

The heat generated by transformers can be considerable and, where it is decided to locate transformers within buildings, these units should be provided with appropriate ventilation to avoid overheating of the transformer and its surrounding space.

6.4.4   Rating

The size of a transformer is determined by its 'rating', which is the product of the voltage and current that a particular unit can handle. In small transformers this rating is given as V·A ('volt-amps') and in larger units as kV·A ('kilovolt-amps'). In the largest transformers the rating is given as MV·A ('megavolt-amps').

Table 6.1 Four-letter codes for cooling systems designation

Letter	Symbol	Description
(a) Oil-immersed transformers		
First letter (identifies internal cooling medium in contact with windings)	O	Mineral oil or insulating liquid with fire point ≤ 300 °C
	K	Insulating liquid with fire point >300 °C
	L	Insulating liquid with no measurable fire point
Second letter (identifies circulation mechanism for internal cooling medium)	N	Natural thermosyphon cooling in windings
	F	Forced oil circulation, but thermosyphon cooling in windings
	D	Forced oil circulation, with oil directed into the windings
Third letter (identifies external cooling medium)	A	Air
	W	Water
Fourth letter (identifies circulation mechanism for external cooling medium)	N	Natural convection
	F	Forced circulation
	Example: ONAN or OFAF	
(b) Dry-type transformers		
First letter (identifies internal cooling medium in contact with windings)	A	Air
	G	Gas
Second letter (identifies circulation mechanism for internal cooling medium)	N	Natural cooling
	F	Forced cooling
Third letter (identifies external cooling medium)	A	Air
	G	Gas
Fourth letter (identifies circulation mechanism for external cooling medium)	N	Natural convection
	F	Forced circulation
	Example: AN or GNAN	

In principal, the transformer has two main determining factors to consider for a given application:

— the rating

— the voltage difference, commonly called the 'voltage ratio'.

To assist in selection and sizing of power distribution transformers, manufacturing and international standards have adopted a table of standardised transformer ratings. The preferred values of rated power (in kV·A), based on ISO 3[17], are stated in BS EN 60076-1[1] as: 100, 125, 160, 200, 250, 315, 400, 500, 630, 800, 1000 etc.

The typical sizes of power transformers found in buildings have primary and secondary currents as defined in Table 6.2.

Table 6.2 Typical sizes of power transformers found in buildings

Rating / kVA	HV current / A	LV current / A	Impedance / %
315	17	455	4.75
500	27	722	4.75
630	34	909	4.75
800	43	1155	4.75
1000	54	1443	4.75
1250	67	1804	5.00
1600	86	2309	5.50
2000	108	2887	6.00
2500	135	3609	6.00
3150	170	4547	6.00
3500	189	5051	6.00

Note: transformer ratio 11000 V/ 416 V; 50 Hz, three-phase, Dyn11

Care in the selection of all associated switchgear and cables must be exercised as larger units can produce extremely high fault levels. These fault levels are a consequence of a short circuit and typically at 3500 kV·A rating, the fault level may be in the region of 75 kA.

The formula for determining the fault level at the secondary terminals of a transformer is:

$$\text{Fault level (kA)} = \frac{\text{Transformer rating (V·A)} \times 100}{\text{Sec. voltage (V)} \times \text{impedance (\%)}} \quad (6.3)$$

A full examination of short circuit currents in 3-phase AC systems, including those in transformer circuits, is described in IEC 60909[18].

6.4.5 Vector groups

The vector group denotes the way in which the windings are connected and the phase position of their respective voltage vectors. It consists of letters identifying the configuration of the phase windings and a number indicating the phase angle between the voltages of the windings.

The convention for defining transformer arrangements is defined in clause 6 in BS EN 60076-1[1]. The star, delta or zigzag connection of a set of phase windings of a three-phase transformer are indicated by the letter 'Y', 'D' or 'Z' for the HV winding and 'y', 'd' and 'z' for the LV winding. If the neutral point of a star-connected or zigzag-connected

winding is brought out, the indication is YN (yn) or ZN (zn) respectively.

To symbolise the phase difference between the primary and secondary windings, a clock number notation is used with 1 o'clock being 30 degrees leading, 2 o'clock being 60 degrees leading etc. The standard UK transformer arrangement is 'Dyn11', i.e a delta-connected HV primary, a star-connected LV secondary brought out to a neutral connection and a phase displacement of 330 degrees leading (30 degrees lagging).

With three-phase AC, the winding connections are identified as follows:

— *delta*: D, d

— *star*: Y, y

— *interconnected star*: Z, z

— *open*: III, iii.

In the case of more than one winding with the same rated voltage, the capital letter is assigned to the winding with the highest rated power; if the power ratings are the same, to the winding which comes first in the order of connections listed above. If the neutral of a winding in star or interconnected star is brought out, the letter symbols are 'YN' or 'ZN', or 'yn' or 'zn', respectively.

To identify the phase angle, the vector of the high-voltage winding is taken as a reference. The number, multiplied by 30° denotes the angle by which the vector of the LV winding lags that of the HV winding. With multi-winding transformers the vector of the HV winding remains the reference; the symbol for this winding comes first, the other symbols follow in descending order according to the windings rated voltages.

Power transformers are delta/star connected three-phase units, normally operating at a vector shift of 30° having the star point connection bought out as the neutral connection. Typically this detail is simply written down as 'Dyn11' and is the preferred UK standard vector group.

This and other examples of transformer notation are shown in Figure 6.4.

Apart from Dyn11 connections, the following are the most likely winding arrangements used in the UK power distribution industry:

— Yyn0: for distribution transformers; the neutral point can be loaded continuously up to 10% of the rated current, or up to 25% of the rated current for a maximum of 1.5 hours (e.g. for connecting arc suppression coils).

— Ynyn0: with compensating winding, used for large system-tie transformers. The neutral point can be loaded continuously with the rated current.

— Ynd5: intended for machine and main transformers in large power stations and transformer stations. The neutral point can be loaded with the rated current. Arc suppression coils can be connected (delta winding dimensioned for the machine voltage).

— Yzn5: for distribution transformers, used up to approx. 250 kV·A for local distribution systems.

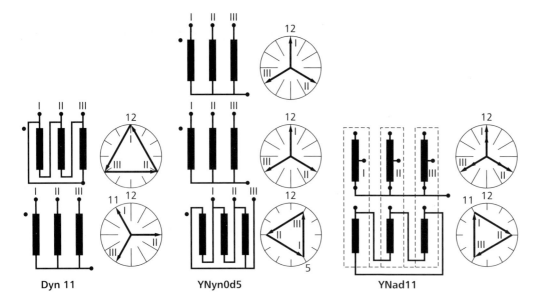

Dyn 11   YNyn0d5   YNad11

Figure 6.4 Transformer notation

The neutral point can be loaded with the rated current.

— Dyn5: for distribution transformers above approx. 315 kV·A, for local and industrial distribution systems. The neutral point can be loaded with the rated current.

— Li0: for single-phase transformers, intended for traction power supply or for three-phase banks with very high voltages and powers.

6.4.6   Primary and secondary electrical supplies

In building services, the main consideration is for HV power distribution transformers, which effectively provide the correct current and voltage requirements for a building or machine. Electricity network distribution from suppliers within the UK standardises all domestic applications to 230 V (single phase) or 400 V (three-phase) and large commercial and industrial users to 11 000 V.

At extremely large industrial plants higher voltages are commonly provided. The 11 000 V standardised figure can vary in a few instances between 6600 and 20 000 volts, depending on the distribution system voltage utilised by the network providers.

For clarity, voltage reference is always expressed as 'no load'. Voltage variation must take into account the magnetising effect of the core in the transformer and this allowance is called the regulation. Therefore, for example, 416 V no-load becomes 400 V on load.

6.4.7   Transformer duty

In specifying a transformer, it is important to consider the duty cycles to which the transformer will be subjected. Guidance on the loading of transformers can be found in IEC 60354[15] and IEC 60905[16].

When a transformer is subjected to load, the high currents in the secondary windings cause heat to be produced in the transformer windings. This, in turn, causes expansion in the transformer. A transformer that is operated at a relatively consistent load will attain a constant temperature. However, in an installation where the loads are operated often for short periods and the transformer is switched between high and low loads at frequent intervals, the transformer windings suffer considerable expansion and, if not sufficiently robust, can fail prematurely.

6.4.8   Location

Technically, the location for transformers plays a large part in the type of unit supplied. Outdoor transformers at ground level tend to be oil-filled units which are the preferred option of the utility companies/network providers. These are pad mounted units with either HV/LV termination chambers or a closely coupled associated HV disconnection device.

Other liquid-filled transformers may be used indoors but careful consideration must be given to the fire/explosive potential for these types of transformers. Due to this potential fire hazard, dry-type transformers are the preferred solution for indoor use. The damp atmosphere in the UK further favours dry cast-resin units and these are now predominately used indoors but can be provided if required with outdoor protection enclosures to IP44.

Where transformers are specified for use in countries other than the UK, particularly where intense direct solar radiation or high summertime temperatures occur, specific facilities such as sun screens may be required. In this situation, the proposed location and anticipated ambient temperature should be advised as part of the information provided to the transformer manufacturer.

6.4.9   Impedance

The short circuit impedance is the sum of the reactance and resistance created in the windings of the transformer after voltage is applied. This impedance is further expressed as 'core losses' (no-load losses) and 'winding losses' (load losses) and is given in watts or kilowatts.

The impedance of transformers is also expressed in percentage values to enable fault level calculations to be carried out. Typical impedance values for various sizes of 11 kV/416 V transformers are as follows:

—   500 kV·A: 4.75%

—   800 kV·A: 4.75%

—   1000 kV·A: 4.75%

—   1250 kV·A: 5.00%

—   1600 kV·A: 5.50%.

6.4.10 Neutral and earthing arrangements

Where star winding arrangements are specified, the centre point should be brought out to a neutral terminal, which should be earthed.

In normal circumstances, the neutral connection will be solidly connected to the system earth but specific circumstances may dictate that the neutral be connected by alternative means. Where there is a requirement to limit short circuit currents, the neutral connection should be connected to earth through a current limiting reactor.

6.4.11 Tappings

Power distribution transformers operate in networks where the high voltage power supplies can be above or below the nominal level. For this reason they are supplied with the facility to vary the nominal primary voltage by a system of switches as in the case of liquid-filled units or through a link board as supplied in dry-type units. This facility is called a 'tap change' and allows the nominal voltage to be adjusted typically by 2.5% and 5% above and below nominal voltage to deliver the correct low voltage requirement.

On power distribution transformers at 11 kV, this setting of the links is undertaken in the de-energised state and is therefore known as 'off-load tap change'. In larger distribution transformers, mainly supplied to utility companies, these tap changers can be motorised and operate without the necessity of de-energising the transformer. This facility is called an 'on-load tap change' and allows distribution companies to switch between differing networks to maximise power usage.

The ability to vary the ratio is important particularly with main transformers. This facility is used for matching the service voltage in the event of load fluctuations, for load distribution or for adjusting active and reactive current in interconnected networks, and for voltage correction with electric furnaces, rectifier stations etc. In the simplest case this is done with the transformer dead, by altering the connection between winding sections with the aid of extra winding terminals, known as 'tappings' (normally +4% or +5%).

Stepwise variation under load is done with a tap changer, usually at the neutral-point end of the HV winding on power transformers and at the series winding on booster transformers and autotransformers.

The tap changer, which connects the respective tappings while under load, basically consists of a load switch and a selector (or alternatively just a selector switch) with or without preselection.

Continuous variation under load requires a special design with moving windings in the form of a rotary or moving-coil transformer.

6.4.12 Terminals and cable boxes

Transformer connections are normally brought out to terminals within cable boxes on the sides of the transformer. The location of the terminals varies between manufacturers and the design of the transformer will determine whether the primary and secondary cable boxes are on the same or opposite sides.

To enable the transformer manufacturer to provide the correct cable boxes, it is important that the designer confirms the following information:

—   size and type of cables

—   number of cables per phase

—   direction of cables entering cable box.

Where single core cables are to be used, insulated gland plates should be specified to avoid eddy currents.

Cable boxes should comply with the requirements of BS 7821[13].

6.4.13 Accessories and tank fittings

The designer should specify the accessories that are required on the transformer. These may include:

—   thermometer

—   temperature contacts (usually 2-stage) for initiating an alarm and tripping the HV supply

—   lifting lugs

—   rollers

—   conservator to permit expansion on oil filled transformers

—   Bucholtz relay for tripping the HV supply on detection of high oil pressure or gassing

—   breather with desiccator

—   oil level indicator

—   filling arrangement

—   drain and sampling plug or valve.

6.4.14 Enclosures

Liquid-filled transformers are manufactured with welded tanks and radiators that have an ingress protection (IP) rating, as defined by BS EN 60529[19], suitable for outdoor location.

Dry-type and cast resin transformers are usually produced without enclosures (i.e. with an IP00 rating) and a suitable enclosure should be specified where there is access to the transformer. This is normally in the form of a louvered steel enclosure with removable access doors. These should incorporate door contacts that are connected such that they trip the HV supply in the event that the access doors are removed while the transformer is live.

6.4.15   Testing

Completed transformers are subject to a number of tests in accordance with the adopted standards for the design. These tests prove design intent, load and voltage withstand capability, thermal ability and operational/functional requirements. Some tests are conducted as routine and others are considered special. Routine tests are normally carried out at the manufacturer's works and are referred to as factory routine tests. Other tests are classified as type tests and special tests and may be conducted elsewhere from the place of manufacture.

6.5   Noise levels and means of noise abatement

If transformers are located in or near residential or other noise-sensitive areas, the noise they produce must be determined so as to assess the need for any countermeasures.

The noise of transformers is defined as the A-weighted sound pressure level measured in dB(A) at a specified measuring surface with a sound level meter, and then converted to a sound power level using the following formula:

$$L_{wa} = L_{pa} + L_{s} \qquad (6.5)$$

where L_{wa} is the A-weighted sound power level (dB), L_{pa} is the A-weighted sound pressure level (dB) and L_{s} is the sound pressure level at the measuring surface (dB).

For transformers with water cooling or fan-less air-cooling, at least six measurements must be taken at a distance of 0.3 m from the surface of the transformer.

The causes and effects of the noise produced by transformers and their cooling systems are so diverse that it is not possible to recommend generally applicable noise abatement measures. Each case must be carefully investigated as necessary.

Possible measures include:

— actions by the transformer manufacturer to reduce airborne and structure-borne noise

— structural measures against airborne noise, e.g. sound-absorbent walls or enclosures

— anti-vibration treatment of the foundations to reduce transmission of structure-borne noise, e.g. spring-mounted supporting structure.

References

1   BS EN 60076: *Power transformers*: Part 1: 1997: *General* (London: British Standards Institution) (1997)

2   BS EN 60044-1: 1999 (IEC 60044-1: 1996): *Instrument transformers. Current transformers* (London: British Standards Institution) (1999)

3   BS EN 60726: 2003: *Dry-type power transformers* (London: British Standards Institution) (2003)

4   BS EN 60742: 1996 and BS 3535-1: 1996: *Isolating transformers and safety isolating transformers. Requirements* (London: British Standards Institution) (1996)

5   BS EN 61378: *Convertor transformers*; Part 1: 1999: *Transformers for industrial applications*; Part 2 (IEC 61378-2: 2001): 2001: *Transformers for HVDC applications* (London: British Standards Institution) (1990, 2001)

6   BS EN 61558: *Safety of power transformers, power supply units and similar devices* (17 parts/sections) (London: British Standards Institution) (various dates)

7   BS 171: 1970: *Specification for power transformers* (London: British Standards Institution) (1970)

8   BS 3941: 1975: *Specification for voltage transformers* (London: British Standards Institution) (1975)

9   BS 6436: 1984: *Specification for ground mounted distribution transformers for cable box or unit substation connection* (London: British Standards Institution) (1984)

10   BS 7452: 1991 (IEC 60989: 1991): *Specification for separating transformers, autotransformers, variable transformers and reactors* (London: British Standards Institution) (1991)

11   BS 7628: 1993: *Specification for combined transformers* (London: British Standards Institution) (1993)

12   BS 7735: 1994 (IEC 60354: 1991): *Guide to loading of oil-immersed power transformers* (London: British Standards Institution) (1993)

13   BS 7821: *Three phase oil-immersed distribution transformers, 50 Hz, from 50 to 2500 kVA with highest voltage for equipment not exceeding 36 kV* (7 parts/sections) (London: British Standards Institution) (various dates)

14   BS 7844: *Three-phase dry-type distribution transformers 50 Hz, from 100 to 2500 kVA with highest voltage for equipment not exceeding 36 kV* (3 parts) (London: British Standards Institution) (various dates)

15   IEC 60354: 1991: *Loading guide for oil-immersed power transformers* (Geneva: International Electrotechnical Commission) (1991)

16   IEC 60905: 1987: *Loading guide for dry-type power transformers* (Geneva: International Electrotechnical Commission) (1987)

17   ISO 3: 1973: *Preferred numbers — Series of preferred numbers* (Geneva: International Standards Organisation) (1973)

18   IEC 60909: *Short-circuit currents in three-phase AC systems*: Part 0: 2001: *Calculation of currents*; Part 1 2002: *Factors for the calculation of short-circuit currents according to IEC 60909-0*; Part 2: 1999: *Electrical equipment — Data for short-circuit current calculations in accordance with IEC 909 (1988)*; Part 3: 2003: *Currents during two separate simultaneous line-to-earth short circuits and partial short-circuit currents flowing through earth*; Part 4: 2000: *Examples for the calculation of short-circuit currents* (Geneva: International Electrotechnical Commission) (1999–2003)

19   BS EN 60529: 1992: *Specification for degrees of protection provided by enclosures (IP code)* (London: British Standards Institution) (1992)

7 Low voltage switchgear and distribution

7.1 Principles of low voltage distribution

7.1.1 Introduction

Many principles have to be considered when defining load centres and types. Empirical and analytical values within the constraints defined by BS 7671[1] and other relevant documentation must be used to establish the locations of the low voltage switchgear and the type of switchgear that would best suit the type of building, space available, environment, system flexibility, capital cost, type of loads and future use of the system.

Locations of the switchgear must offer solutions to all the above points and still offer a viable solution to serving the loads that will exist throughout a building. The design engineer will look for switchgear positions that limit the use of expensive and unnecessary long cable runs within the system and offer convenience for future maintenance.

For low rise buildings with heavy point loads such as industrial premises it is usual to place the switchgear local to substations.

Multi-storey buildings are usually served via switchrooms located at lower levels, i.e. basements, car parks etc. with service risers to each floor. It is usual to find individual tenant and landlord risers in commercial retail and lettable units. These often incorporate facilities to allow sub-metering. It is also common to find multi-storey buildings with substations located at low level and high level. This system when used in conjunction with high level plantrooms offers a great deal of flexibility.

For all situations where switchgear, distribution boards, control equipment etc are used to satisfy load centres as a general principle it is more economic to position the switchgear as near as possible to the loads. When the switchgear and distribution equipment can be located close to the load it reduces sub main cable sizes, trunking cable tray and conduit runs.

7.1.2 Circuit description

It is necessary for every installation to be divided into individually protected circuits to avoid danger and minimise inconvenience in the event of a fault.

In addition, circuit division will aid safe operation, inspection, testing and maintenance.

Various attributes of a circuit are described by the designated area that it serves. Three categories separate these:

— *Main*: the circuit that connects the main switchboard to any sub-main switchboard.

— *Sub-main*: the circuit that connects any of the sub-main switchboards to any one distribution board.

— *Final circuit*: a circuit that connects directly to current consuming equipment, or other points of connection such as a socket outlet for the connection of such electrical loads. Final circuits are divided into two types: *ring circuits* and *radial circuits*.

The number of final circuits required, and the number of points that the final circuit supplies must satisfy the requirements detailed in BS 7671[1]. These criteria are as follows:

— over-current protection (BS 7671, section 43)

— isolation and switching (BS 7671, section 46)

— current-carrying capacities of conductors (BS 7671, section 52).

7.2 Low voltage equipment

7.2.1 Switchgear components

Switchgear used in low voltage (LV) distribution incorporates many components, consisting primarily of spring action contacts, interlocks, arc controllers and cable connections, frames and enclosures and protective devices

The switchgear is used to open and close the circuit. Main switchgear will usually offer protection to the circuit against over-voltage, under-voltage, overloads, fault currents and earth leakage. Protective devices can be selected to offer all of these facilities or a combination of them, as appropriate.

Simple switches or switchgear can be defined as mechanical devices that are non-automatic, and used for opening and closing circuits operating under normal load conditions. When used to open and close the circuit under normal conditions these switches must switch with a minimum of arcing. Such devices are hand operated, air break with a fast make and break action.

The isolator is a device that isolates a circuit. As such, the isolator is a form of knife switch or withdrawable link that, when operated or withdrawn, forms a break in the circuit from the source of supply to the load. Isolators do not provide fast make and break facilities and, because of this, give rise to severe arcing when used to break circuits under load. They should therefore only be provided to isolate supplies that have been opened on-load by other forms of switchgear.

A contactor is an electromagnetic or electro-pneumatic device used to open and close circuits under load conditions. Contactors are used to control motors or other loads and are used for other applications, i.e. remote and/or automatic control. The contactor may be controlled in a number of ways through energising an electromagnetic coil. The circuit controlling the coil may incorporate stop, start, inch and auxiliary switches.

Figure 7.1 shows the application of a contactor as a direct-on-line motor starter. Direct-on-line starting is generally only used on motors rated up to a maximum load of 7.5 kW.

The coil is energised by pushing the start button, when a 400 V supply will be connected across the coil causing it to induce a magnetic field. The coil pulls in the load contacts L1, L2, L3 and the holding contact C1. The start button is spring-loaded and the operator will press the button momentarily to initiate the start. The holding contacts are in series with the stop button and will allow some current to continue to flow to the coil and hence keep the contactor energised. The contactor will be tripped by pressing either the stop button or the trip button, or if overload conditions are detected. Any of these will break the supply to one side of the coil, the contactor will de-energise and, through spring or gravity action, the contacts will open. The contactor will not re-close automatically as contacts C1 have also opened, this gives the contactor a 'no-volts' release characteristic.

For larger loads, contactors may also be controlled via other forms of circuit wiring, such as star-delta, automatic transformer and inverter control.

7.2.2   Forms of separation

Various recognised standards of switchgear construction exist and these are detailed in BS EN 60439-1[2].

Placing equipment in a basic enclosure will provide a degree of separation, as the door can be shut to prevent accidental contact between live parts. However in most applications there are benefits to be gained by providing additional separation between components.

BS EN 60439-1[2] defines four forms of separation that represent the construction arrangements universally available in switchroom assemblies. The forms cover the separation of busbars, functional units and incoming and outgoing terminations.

These are summarised as follows:

(a)   *Form 1*: an assembly fashioned to provide protection against contact with internal live parts, although there is no internal separation. The following conditions apply:

—   busbars are not separated from functional units

—   functional units are not separated from other functional units

—   functional units are not separated from any incoming or outgoing terminations

—   busbars are not separated from any incoming or outgoing terminations.

(b)   *Form 2*: typically the overall enclosed assembly provides protection against contact between internal live parts and internal separation of busbars from functional units.

Form 2a: as form 2, terminals are not separated from the busbars or each other.

This guide is subdivided into type 1 and 2 to differentiate between the use of insulated coverings and the use of rigid barriers to achieve the separation:

Form 2 Type 1: as form 2, the busbars are separated with the use of insulated coverings. The busbars are separated from the terminals, but not from functional units or each other.

Form 2 Type 2: as typical form 2 arrangement. The busbars are separated with the use of metallic or non-metallic rigid barriers. The busbars are separated from the terminals, but not from functional units or each other.

(c)   *Form 3*: the typical arrangement is as follows:

—   busbars are separated from the functional units

—   functional units are separated from other functional units

—   functional units are separated from incoming and outgoing terminations

—   busbars are not separated from incoming and outgoing terminations.

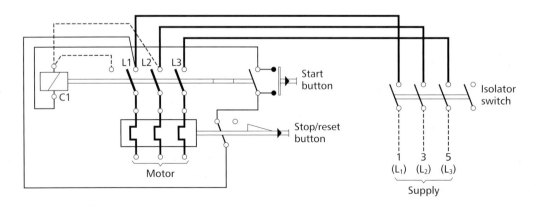

Figure 7.1 Contactor used as a direct-on-line motor starter

Form 3a: as typical form 3 arrangement. The terminals are not separated from the busbars or each other. This guide is then subdivided into type 1 and 2 to differentiate between the use of insulated coverings and the use of rigid barriers to achieve the separation.

Form 3b, Type 1: as typical form 3 arrangement. The busbars are separated with the use of insulated coverings. The busbars are separated from the terminals, but not from each other.

Form 3b, Type 2: as typical form 3 arrangement. The busbars are separated with the use of metallic or non-metallic rigid barriers. The busbars are separated from the terminals, but not from each other.

(d)   *Form 4*: The typical arrangement is:

— busbars are separated from the functional units

— functional units are separated from each other

— terminations to functional units are separated from each other

— functional units have a separate neutral connection with a means of disconnection.

Form 4a, Type 1: as typical form 4 arrangement. Busbars are insulated with coverings. Cables are terminated within the same unit as the associated functional unit. Cables maybe glanded elsewhere.

Form 4a, Type 2: as typical form 4 arrangement. Busbar separation is achieved with the use of rigid barriers or partitions. Cables are terminated within the same unit as the associated functional unit. The cables maybe glanded elsewhere, i.e. a cable chamber.

Form 4a, Type 3: as typical form 4 arrangement. Busbars are insulated with coverings. Cables are terminated within the same unit as the associated functional unit. Terminations for each of the functional units have facility for integral glanding.

Form 4b, Type 4: as typical form 4 arrangement. Busbars are insulated with coverings. Terminals are external to the functional unit, and insulated coverings provide separation. Cables maybe glanded elsewhere, i.e. a common cable chamber.

Form 4b, Type 5 (terminal boots): as typical form 4 arrangement. Busbar separation is achieved with the use of rigid barriers or partitions. Terminals are external to the functional unit, and insulated coverings provide separation. Cables maybe glanded elsewhere, i.e. a common cable chamber.

Form 4b, Type 6: as typical form 4 arrangement. Rigid barriers or partitions provide total separation.

Terminals are external to the functional unit, and rigid barriers provide compartmentation. Cables maybe glanded elsewhere, i.e. a common cable chamber.

Form 4b, Type 7: as typical form 4 arrangement. Rigid barriers or partitions provide total separation. Terminals are external to the functional

unit, and rigid barriers provide compartmentation, complete with integral glanding.

The forms of construction guidance has been written around various methods and means of preventing a fault in a single circuit being transferred to adjacent circuits. In general terms the higher the form rating the higher the degree of protection.

It should be noted that a panel constructed to the requirements of a particular form does not necessarily mean that a live panel is safe to work on or operate without being fully electrically isolated.

In general terms the higher the form of separation the more costly is the switchgear assembly. It is therefore important that the specific application be carefully considered prior to specifying the switchgear. For large 'mission critical' applications such as hospitals, large financial organisations and military applications, it may be prudent to specify a derivative of form 4 construction. This will provide protection against single circuit failure in a cubicle spreading to other circuits. Alternatively, for a development where non-mission critical activities occur such as light retail or large domestic, it would be prudent to specify a lower form of switchgear separation. The main application for forms 1 and 2 would for motor control centres serving non-critical mechanical plant. The standard form of separation for a standard miniature circuit breaker (MCB) distribution board or consumer unit is considered to be form 1.

7.2.2   Distribution boards

Distribution boards will usually be used at subdistribution points to provide a means where final circuits may be connected to the loads through fuses or circuit breakers. Distribution boards should be located as close to the load centres as possible to reduce the amount of final circuit wiring and cable containment systems used, i.e. trunking, conduit etc. Distribution boards will usually include high rupturing capacity (HRC) fuses in old installations, or more commonly in modern installations, miniature circuit breakers (MCBs). They may be used for a variety of applications and may be single or three-phase configuration dedicated lighting boards, power boards, or multi-service boards. Depending on the application the board may be purchased with a main isolator switch rated at a desired current rating.

7.2.3   Circuit protective devices

7.2.3.1   Fuses

Fuses are the simplest form of circuit protection. The fuse is made up of a number of components. A fuse element, incorporated within a carrier, is designed to melt when too much current is passed, hence breaking or opening the circuit under overload or fault conditions.

Fuse carriers are located within distribution boards or fuse switches and are designed to enable the carrier to be withdrawn from the circuit without having to touch live parts of the distribution board.

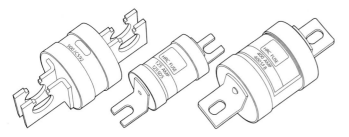

Figure 7.2 Typical high rupturing capacity (HRC) fuses

Rewireable fuses, whilst still retained in many old installations, are not generally used in new installations. The most commonly used fuse is the high rupturing capacity (HRC) fuse as shown in Figure 7.2. The quartz filling will extinguish any arcs that are formed under fault conditions by filling the gap created between the two ends of the fuse element with a high resistance powder.

A benefit of fuses is that they withstand much higher fault levels than other electromechanical forms of protection.

Fuses are not interchangeable; the connection tags are spaced differently depending on the rating of the fuse. This prevents increased fuse sizes being installed within a reduced capacity frame.

Fuses are commonly used and are generally considered to be reliable in operation.

7.2.4   Circuit breakers

Circuit breakers are commonly used today for protecting circuits on HV and LV circuits. The circuit breaker is a mechanical device capable of opening and closing circuits under normal and fault conditions. For low current, low voltage applications miniature circuit breakers (MCBs) may be used to protect final circuits. Characteristics of MCBs are given in BS 7671[1] and manufacturers' information.

MCBs are manufactured to recognised standards. These standards provide three different categories of performance depending on the application. These are as follows:

— *Type B*: used where resistive loads are present such as tungsten lighting or electric heating is present.

— *Type C*: used where a general mixture of light inductive and resistive loads are present, such a light commercial.

— *Type D*: used where there is a strong presence of inductive loads such as motors and switched mode power supplies.

Each breaker type should be selected following a detailed assessment of the load types.

Earth fault currents are detected and controlled by residual current circuit breakers, known as 'residual current devices' (RCDs). RCDs consist of three windings on a ring core. Two of the windings are connected to the live and neutral respectively, and are wound onto the core equal and opposite. The third winding is connected to the tripping coil. Under normal conditions the magnetic flux between the phase and neutral is zero, hence no current

flows. However, when an earth fault occurs, the current flowing in the neutral will be less than the current flowing in the phase conductor, causing an imbalance in the two windings which induces a current in the third tripping coil winding. When the current has reached the magnitude at which the RCD is set to trip (i.e. 30–100 mA) then the tripping coil will open the circuit.

An RCD should also incorporate a test button connected to the internal circuitry which, when activated, will test the circuit to ensure correct operation. However, this tests only the mechanical elements of the device, not the electrical. The electrical elements should also be subjected to a trip test using suitable test equipment.

It is possible to combine an RCD and an MCB to provide a device known as a 'residual current breaker with overcurrent protection' (RCBO). This device would generally be located within the distribution board. Consideration should be given to the fact that some manufacturer's units occupy two outgoing ways on a distribution board whereas units from other manufacturers only occupy a single outgoing way. These devices offer the benefits of a compact solution to providing circuit protection with the added benefits of earth leakage protection.

HRC fuses offer some benefits over modern day miniature circuit breakers (MCBs) or moulded case circuit breakers (MCCBs), such as high fault withstand ability, silent operation and simple interchangeability. However MCBs or MCCBs do offer other advantages such as simple identification of which device has operated, the ability to automate the device and, on larger devices, the ability to alter the time–current characteristic to assist with discrimination.

For these reasons, it is important to consider the particular requirements of each project prior to selecting the protective equipment.

7.2.5   Alternative means of serving load centres

Load centres may be served in a number of ways. Most supplies are simply taken radially via HV transformers, sub-distribution boards and distribution boards to the final circuits. Alternatively, there are many applications where the load centre must be protected from losing its supply, e.g. hospitals, computer centres, production processes etc. For these applications, security of supply to the load centre can be established by providing either uninterruptible supplies or standby generator back-up.

Standby or emergency generators can be sized to provide the entire building load or a part of the full load. Where generators are required to take the full load of the building automatically they are interlocked with the mains supply. As such, they sense the loss of supply and initiate the prime-mover (usually a standby generator); once the prime mover is at the correct speed the generator will be switched on-line, so taking the full load of the building. (Whether it provides full or part load, the generator is interlocked with the main supply.)

Constraints such as capital cost, space etc. may limit the amount and size of generators that may be installed within a building and therefore the emergency supply available may not be capable of taking the full load of a particular building. Under these circumstances the loads must be divided into 'essential' and 'non-essential' loads and the essential loads only connected to the generator.

In a hospital the essential loads will be the equipment which is critical for patient systems and staff safety. This may include nurse call systems, communications systems, medical gas supplies, socket outlets for life-support equipment, lifts, theatre suites and fire alarm systems. The non-essential circuits will supply cleaners' socket outlets, wall lights and other general items that are not considered essential to maintain the services required by the staff. Under this system it is common to split the lighting circuits such that 50% of the lighting is regarded as an essential load and 50% as non-essential.

7.2.6   Supply voltages

Low voltage is defined as 50–1000 V AC or 0–1500 V DC. The majority of distribution networks installed in buildings operate at 400 V three-phase and/or 230 V single phase. In large and industrial buildings it is common for high voltage supplies (typically 11 000 V) to be provided to local substations where it is transformed down to 400 V for use by the consumer. The 400 V supply is then distributed to various load centres from which 400 V three phase and 230 V single phase supplies are provided.

Where three-phase distribution is selected, it is important that the designer considers the issue of load balancing between the three phases.

Figure 7.3 shows a typical factory arrangement in which the incoming HV supply is transformed down to 400 volts and distributed to three-phase and single phase circuits.

7.3   Equipment selection

7.3.1   Switchgear

Switchgear used for LV distribution incorporates a number of common components. The purpose of the switchgear is to control the circuit under normal and fault conditions. The switchgear specified and installed must be capable of making and breaking circuit loads via switch contacts or in some cases the tripping action of the switchgear.

Most non-automatic switchgear will have some form of manually operated handle to open and close the switch. The handle will usually be spring loaded to ensure quick make and break facility. Switchgear will also have contacts that are capable of withstanding the load of the circuit and the fault current that may exist within the circuit.

Some form of arc quenching should be incorporated within the switchgear. This will extinguish the arc formed within the switchgear when opening a circuit under load conditions (and fault conditions if the unit has automatic overload protection).

Environmental factors will greatly influence the choice of switchgear. Most switchgear, from large industrial to domestic uses, is usually robust and mechanically strong. However, careful consideration will need to be given to the positioning of the switchgear. Unless the switchgear is specifically designed for arduous conditions, it must be located in a moisture and dust free atmosphere. Under normal situations the switchgear will be placed in a convenient cupboard, switchroom or plantroom, located as close to the load centre as possible.

In many cases it is necessary to provide supplies to equipment in hazardous or corrosive environments. In these circumstances it is more economic to employ standard switchgear placed close to, but outside, the

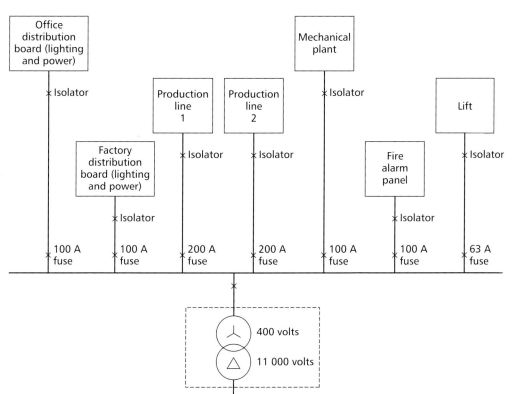

Figure 7.3 Typical supply arrangements for a factory

hazardous environment than to install specially manufactured switchgear within the hazard area itself.

7.3.2   Ingress protection (IP)

An important consideration in the selection of electrical equipment is the extent to which the device is protected from ingress by liquids or solids. The system of classification is fully described in BS EN 60529[3].

The level of protection against the intrusion of solid and liquid particles into the electrical equipment is specified by the ingress protection (IP) rating. An IP rating has two digits; the first digit represents the degree of protection against the ingress of solids and the second represents the degree of protection against the ingress of liquids.

The degrees of protection against ingress of solids and liquids are shown in Table 7.1.

7.3.3   System selection

For high voltage applications the available options for the selection of switchgear is limited. However for LV switchgear there are numerous manufacturers offering a wide range of options.

Selection of equipment must be based on a number of considerations, including:

—   type and number of loads to be served

—   space available to install the switchgear

—   capital cost of equipment

—   access to cable containment (i.e. trunking, trays, conduits or cable runs)

—   degree of isolation consistant with regular maintenance and alterations, both when live and dead.

In many instances the choice is dictated by constraints that may be inherent with the type of building, space allocation, available capital etc. However when choosing the type of switchgear for a particular application the designer must fully consider the above and, equally important, what may be required in the future.

Certain loads and environments will dictate the use of specific switchgear. The designer will always attempt to place the distribution board, switchgear, contactors etc. as close to the load centre as possible. The following examples show some typical arrangements for low voltage switchgear for different applications.

Example 7.1: Hospital

Figure 7.4[4] shows a typical hospital with theatres, acute wards etc. all served via essential and non-essential distribution boards.

A 400 V supply is distributed through a busbar chamber to the LV switchgear or via a centralised modular form panel (it is usual to incorporate the power factor correction equipment at this point). From the substation LV switchgear, a typical installation would feed out to the load centres via suitable cables run in voids, ducts, basements etc. to feed further switchgear, MCCBs or sub-distribution boards.

Emergency generators are usually located close to the building's main LV switchgear panel. The emergency generator will provide a 400 V, 50 Hz electrical supply whenever the normal supply is interrupted. As such the generator will be sized to provide a sufficient electrical supply to all essential supplies throughout the building.

Establishing critical loads is vital to the satisfactory operation of the building. During the design of any buildings that will utilise emergency generators to provide essential supplies, it must be remembered that any emergency lighting, work area supplies, kitchen refrigeration etc. should be on the 'essential' supply.

Example 7.2: Retail store

In this case the incoming electrical mains supply is shown entering the building via the back of the store, see Figure

Table 7.1 Ingress protection (IP) ratings

First digit	Level of protection against ingress of solid objects	Second digit	Level of protection against ingress of liquids
0	No protection	0	No protection
1	Protected against large solid objects of 50 mm diameter or greater (e.g. hands)	1	Protected against vertically falling drops of water
2	Protected against small solid objects up to 12.5 mm diameter or greater (e.g. fingers)	2	Protection against dripping liquid falling vertically when enclosure is tilted at an angle up to 15° from vertical
3	Protected against solid objects up to 2.5 mm diameter or greater (e.g. tools)	3	Protection against water sprayed at any angle up to 60° from vertical
4	Protected against solid objects up to 1.0 mm diameter diameter or greater (e.g. wire)	4	Protection against splashing from all directions
5	Protected against ingress of dust in an amount sufficient to interfere with the satisfactory operation of the equipment or impair safety	5	Protection against water jets from any direction
6	Totally protected against ingress of dust.	6	Protection against strong water jets from all direction
—		7	Protected against temporary immersion
—		8	Protected against continuous immersion

Note: when a digit is not required it is replaced by 'x' or 'xx' if neither digit is required

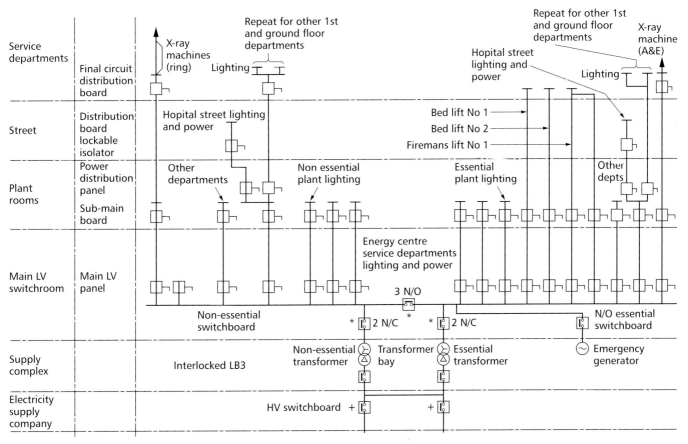

Figure 7.4 Typical distribution for a hospital (reproduced from NHS Estates Health Technical Memorandum HTM 2007[4] (Crown copyright))

7.5. The supply would most likely terminate at the incoming mains positions with a supply taken across to the electrical switchroom. The electrical supply would then terminate in a switchfuse of sufficient size to control the store load. From the main switchfuse other switchfuses would provide supplies to mechanical plant, store signs, distribution boards etc.

7.4 Protection and control/metering

7.4.1 Protection principles

Circuits need to be protected against faults such as overloading, short circuits, under-voltage and earth leakage.

Prolonged overloads may result in heat generated within the circuit, which may lead to a fire within the switchgear or cables. This can quickly spread to adjacent circuits or equipment. Excessive currents can occur when short circuit conditions apply, hence arcs or flashes may again result in unbalanced conditions, insulation breakdown and excessive current, all endangering people or livestock.

Under-voltage conditions will especially affect electric motors, which may slow down, stall or stop altogether. Hence the circuit must be protected against a number of conditions. The designer of a system must also ensure that the protective devices offer sufficient discrimination (i.e. if a circuit has a number of protective devices that are connected in series, the device nearest the fault should operate to clear the fault leaving all other healthy

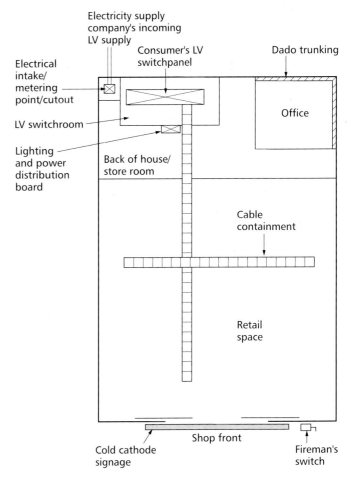

Figure 7.5 Typical distribution for a retail store

circuits operational). Figure 7.6 shows a typical circuit feeding through various protective devices.

For the circuit shown in Figure 7.6, if a fault should occur at point 1, the 10 A final lighting circuit fuse should operate thereby clearing the fault and leaving the remaining circuit operating. A fault at point 2 should be cleared by the 30 A fuse which again leaves the majority of the system operational. A fault at point 3 should operate the 100 A fuse which would result in the loss of a substantial amount of the system but would, however, still leave the 100 A and 63 A circuits operational. A fault at point 4 should be cleared by the 300 A fuse and would result in the loss of all circuits.

Figure 7.6 suggests that the fuse closest to the fault condition would clear the fault thereby avoiding the operation of larger fuses down the line and thus preventing other parts of the circuit from being affected.

However, different protective devices (e.g. MCBs, HRC fuses, semi-enclosed fuses etc.) have different time–current characteristics (i.e. the ability to clear faults of various current magnitudes within various times) and this must be borne in mind when using different types of device in combination.

7.4.2   Metering

Under most conditions the local supply company will connect metering equipment to the incoming supply. For supplies up to approximately 50 kV·A the supply company's meters will be connected direct; for supplies above 50 kV·A the supply will be metered via current transformers.

For consumers above 50 kV·A the local supply company will most probably provide a maximum demand meter. The maximum demand meter will be set to monitor both peak local demand and kW·h units used. An additional

charge will be levied for electrical consumption above a level agreed with the consumer. Various metering configurations and arrangements are available and the various electricity supply companies should be consulted.

Low voltage switchgear should offer adequate discrimination, see section 7.4.1. The designer will need to calculate the fault current, voltage drop, shock protection and load current for each section of the installation. On large installations the designer will usually carry out the installation from incoming MV supply, through MV switchgear, transformer, LV switchgear protected via BS 88 fuses, MCCBs or air circuit breakers (ACBs), through to centrally located switchgear and finally to the distribution board and, in turn, the final circuits.

Where installations are provided with emergency standby generators, double busbar systems etc., it may be necessary to provide a bypass facility. This will provide the facility to bypass the emergency generator, thereby completely isolating the generator from any supply and thus allowing work to take place on the generator. It is essential that the bypass cubicle be interlocked with the generator control panel so that the generator set cannot be switched to the normal mains supply when not synchronised. The interlocking of the two systems is normally carried out using mechanical interlocks. The key isolators will operate via mechanical interlocks that will only allow switchgear to be opened for inspection with the correct keys in place within the panel.

Part L2 of the Building Regulations[5] requires that reasonable provision be made to enable at least 90% of the estimated annual energy consumption of each fuel to be accounted for. This can be achieved by various means but Building Regulations Approved Document L2[6] suggests the size of plant for which separate metering would be considered reasonable, see Table 7.2.

Table 7.2 Sizes of plant for which separate metering is recommended[6]

Plant item	Rated power / kW
Electric humidifiers	10
Motor control centres providing power to pumps and fans	10
Final electrical distribution boards	50

7.4.3   Physical arrangements for LV switchgear

LV switchgear used for large projects will inevitably require a considerable amount of space. Whilst it is always more economic to place the switchgear local to the loads, consideration should be given to the value of usable floor areas. Switchroom sizes must be small enough therefore to be favourable to the designer and to the client.

Switchgear and cables can be provided in various configurations, typically:

—   *cables*: top access or bottom access

—   *switchgear*: front access or rear access

Each building will have differing requirements depending on the space available.

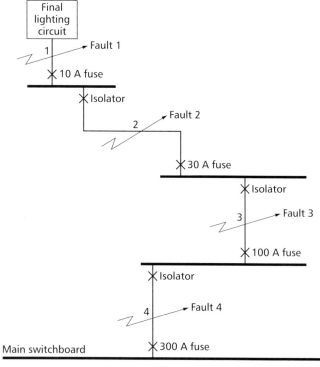

Figure 7.6 Typical protection of a circuit

Note that bottom access switchgear will require cable trenches or pits located beneath the switchgear which will necessitate early coordination with the architect and structural engineer. Where switchgear is mounted low down in bottom entry switch panels, consideration should be given during the design to the location of glanding plates and the distance between glanding plates and the terminals of switchgear. This can be particularly critical where large (>120 mm²) or solid core, cables are installed due to the difficulty in bending the cables in a potentially confined space.

For switchgear located near to load centres, it will be necessary to provide sufficient ducts, tray work, ladder-rack etc., to enable large cables to be installed to and from the switchgear.

Different switchgear arrangements offer access to the busbars from a variety of positions. The maintenance engineer requires safe access to operate any switchgear. The installer requires safe all round access, hence sufficient access must be left at all times for day-to-day use of the switchgear and access for maintenance, testing and future expansion of the system.

Figure 7.7[4] shows examples of space requirements and builders work required when installing LV switchgear. It will be noted that the trench provides flexibility for bringing cables in and out of the LV switchgear.

Typically, builders' work would also cover the provision of adequate drainage and waterproofing of the trenches. Consideration should be given to the water table.

Other points for consideration concerning the layout and contents of the switchroom include the following:

— Trench chequer plates should be cut to suit the equipment and cables. The chequer plates should not extend under switchgear.

— A non-dusting screed should be laid with the final floor level to be minimum 100 mm higher than the outside ground level.

— Panic bolts should be fitted to external doors, preferably two doors per switchroom. Access to the switchroom must meet fire escape requirements and permit access of both personnel and equipment.

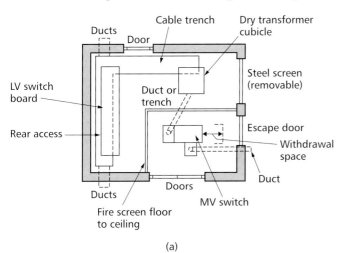

(a)

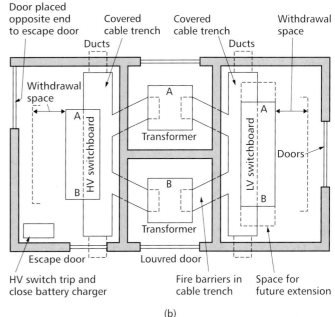

(b)

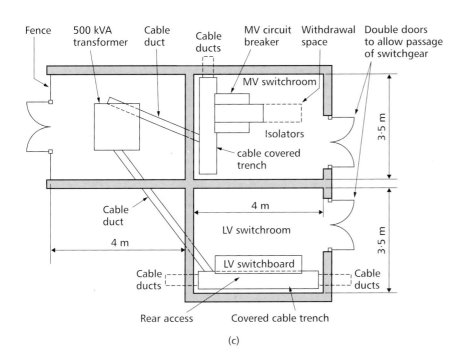

(c)

Figure 7.7 Space requirements and builders' work required when installing typical switchgear; (a) substation with single HV switch, dry transformer and LV switchboard, (b) substation with HV switchboard, two transformers and LV cubicle-type switchboard, (c) small substation with indoor transformer and HV ring main unit switchgear (reproduced from NHS Estates Health Technical Memorandum HTM 2007[4] (Crown copyright))

— Emergency lighting should be provided.

— The switchroom should be designed to achieve a minimum fire resistance of 1 hour.

— The switchroom should be lockable to prevent unauthorised access.

— Rubber mats running the entire length of the switchroom should be provided.

— Relevant safety information should be displayed.

— Resuscitation procedure charts should be displayed.

— A framed, up-to-date schematic of the equipment should be displayed.

7.5 Testing and commissioning

7.5.1 Testing

Testing should be undertaken to satisfy several purposes:

— to verify the adequacy of the design

— to ensure that the installation is sound

— to see that the installation complies with the specification

— to confirm that the installation functions correctly and safely.

It is essential that a thorough visual inspection of the equipment and installation be carried out prior to commencing with the formal testing process. This process is described in BS 7671[1] and is referred to as 'initial verification'.

It is particularly important to remember that electronic components are susceptible to damage from the voltages applied during the test process and therefore it is essential when carrying out such insulation resistance tests to ensure that electronic components such as switched mode power supplies and lighting control gear is disconnected.

Testing for compliance with the specification particularly relates to factory built assemblies or to items of plant which are built at the manufacturer's works and delivered to site fully assembled. Such tests may cover a broad spectrum including safety, function and performance tests. All tests undertaken at the manufacturer's works should be carried out in accordance with the appropriate British Standards. In certain cases the designer may specify that the equipment, e.g. switchgear and distribution power transformers, are 'type tested' at the manufacturer's works. However, standard routine tests will need to be carried out on site to ensure that the switchgear has not been damaged in transit or incorrectly re-assembled. All on-site tests must be performed by qualified and experienced commissioning engineers with experience of the type of equipment which has been installed.

Appropriate test certificates should be retained in the operating and maintenance manuals produced for the building.

7.5.2 Commissioning

When the installation is complete, commissioning will start and the switchgear will be energised. Commissioning requires that the installation work has been correctly carried out and that it has been completed. A thorough check will need to be carried out to ensure that this is so. Pre-commissioning checks should be the responsibility of the installer. However, if applicable, the design engineer is normally present to witness parts of the works under test.

Switchrooms must be complete, weatherproof and without other trades, particularly 'wet trades' working adjacent to HV plant under test. The safety officer for the building or site must be made aware of the scheduled dates for testing and commissioning in order that the necessary safety procedures are in place and that other persons are made aware that testing and commissioning will be taking place.

Commissioning can be complicated but it is essential that it is done properly with adequate numbers of skilled personnel present, both to carry out the commissioning and to witness the respective tests. Ideally, the maintenance staff should be present at this stage to learn about the systems and to provide additional supervision. It is essential that adequate time is allowed in the construction programme to allow commissioning to be carried out properly and in an orderly fashion.

References

1   BS 7671: 2001: *Requirements for electrical installations. IEE Wiring Regulations. Sixteenth edition* (London: British Standards Institution) (2001)

2   BS EN 60439-1: 1999: *Specification for low-voltage switchgear and control gear assemblies. Type-tested and partially type-tested assemblies* (London: British Standards Institution) (1999)

3   BS EN 60529: 1992: *Specification for degrees of protection provided by enclosures (IP code)* (London: British Standards Institution) (1992)

4   *Electrical services supply and distribution — Design considerations* NHS Estates Health Technical Memorandum HTM 2007 (London: Her Majesty's Stationery Office) (1993)

5   Building Regulations 2000 Statutory Instruments 2000 No. 2532 (London: The Stationery Office) (2000)

6   *Conservation of fuel and power* Building Regulations 2000 Approved Document L1/L2 (London: The Stationery Office) (2000)

8 Building wiring systems

8.1 Introduction

This section describes the methods of distributing electricity in buildings. The different types of cables and wiring systems are described in detail, together with advice on determining which types of system are appropriate for the various types of installation.

The specific requirements for fire detection and other life safety systems are covered only with respect to the choice and installation of the cables and the relationship to the electrical wiring.

Frequent reference is made to BS 7671[1] and other British Standards but the text of these documents has not been duplicated here.

8.2 Symbols, abbreviations and definitions

8.2.1 Cable construction abbreviations

Abbreviations used for common types of cable construction are as follows.

EPR	Ethylene propylene rubber
HOFR	Heat and oil resistant and flame retardant
LSF	Low smoke and fume (*Note*: this is not a British Standard definition)
LS0H	Low smoke, zero halogen (*Note*: this is not a British Standard definition)
MI	Mineral insulated
MICC	Mineral insulated copper covered
MIND	Mass impregnated non-draining
PVC	Polyvinyl chloride
SWA	Steel or single wire armoured
XLPE	Cross-linked polyethylene

8.2.2 Definitions

Wherever possible the definitions used in this section are in accordance with BS 7671[1]. These definitions are not repeated herein. Other commonly used terms are defined as follows:

— *high voltage*: any voltage exceeding 1000 V AC

— *containment system*: any enclosed system specifically used for cables and wires; this includes conduit and trunking systems but not support systems such as cable trays and ladders

8.3 Circuit types

8.3.1 Voltage bands, categories and circuit segregation

All editions of the IEE Wiring Regulations[1] (hereafter referred to as 'the Wiring Regulations') have included regulations about which types of circuits can be incorporated into the same cable or containment system.

Previous editions of the Wiring Regulations defined three categories of circuit:

(a) *Category 1*: circuits (other than fire alarm or emergency lighting circuits) operating at low voltage and supplied directly from a mains supply system.

(b) *Category 2*: with the exception of fire alarm and emergency lighting circuits, all extra low voltage circuits; and telecommunications circuits which are not supplied directly from a mains supply system.

(c) *Category 3*: fire alarm and emergency lighting circuits.

The regulations basically stated that category 2 circuits could only be installed with category 1 circuits if the insulation was rated for the highest voltage present and that category 3 circuits could not be installed with category 1 circuits under any circumstances. In practice this means that a telephone circuit would have to be installed using the same type of wiring as a lighting or power circuit, rather than the much smaller types normally used.

These definitions are no longer included in BS 7671[1] but are still often referred to by both designers and installation contractors.

The current edition of BS 7671 defines two voltage bands:

(a) Band I covers:

— installations where protection against electric shock is provided under certain conditions by the value of voltage

— installations where the voltage is limited for operational reasons (e.g. telecommunications, signalling, bell, control and alarm installations); extra-low voltage (ELV) will normally fall within voltage band I.

(b) Band II contains the voltages for supplies to household, and most commercial and industrial installations. Low voltage (LV) will normally fall within voltage band II. Band II voltages do not exceed 1000 V AC RMS or 1500 V DC.

Regulation 528-01 details the current requirements for segregation but the basic requirement is that band I circuits must have insulation rated for the highest voltage present in the band II circuits. Fire alarm and emergency lighting circuits have to be segregated from all other cables, and from each other, as defined in the relevant British Standards and telecommunications circuits also have to be segregated in accordance with the appropriate British Standard.

8.3.2   Sub-mains

The term sub-main, sometimes written without the hyphen, is commonly used to describe the circuit connecting a main switchboard or main incomer to the distribution boards. The closest definition in BS 7671[1] is that for 'distribution circuit'.

These circuits may be installed using single or multicore cables or even busbars in a large installation.

8.3.3   Final circuits

BS 7671[1] defines a final circuit as a circuit connected directly to current-using equipment, or to a socket-outlet or socket-outlets or other outlet points for the connection of such equipment. Final circuits are sometimes referred to as final sub-circuits.

Guidance Note 1[2] gives details for assessing the diversity to be used when calculating the number of final circuits required and also gives details of final circuits for household installations.

8.3.3.1   Ring final circuits

Ring circuits, commonly known as 'ring mains', are used almost universally in the UK to supply socket-outlets but are uncommon in countries that do not use a version of the Wiring Regulations[1]. A sketch of the principle is shown in Figure 8.1.

The concept was introduced shortly after World War II, in conjunction with the BS 1363[3] socket-outlet and 'square pin' 13 A fused plug. Overload protection for the ring circuit is provided by a 30 A fuse or 32 A circuit breaker

and overload and short circuit protection for the flexible cable is provided by a fuse in the plug. The unusual rating of 13 amps was chosen to allow the connection of 3 kW electric heaters.

Previous editions of the Wiring Regulations included details of typical ring circuits but this has now been moved to Guidance Note 1[2]. This states that the maximum area covered by each ring main should not exceed 100 m² for household installations. Ring circuits are normally wired in 2.5 mm² conductors but it may be necessary to increase this to 4 mm² where the circuits are unusually long.

Spurs may be connected to a ring main as shown in Figure 8.2. Unfused spurs may only supply one socket-outlet (single or twin) or one permanently connected item of equipment. The maximum current that can be taken from a twin socket-outlet is 26 A and this is below the maximum rating of a 2.5 mm² cable. Permanently connected equipment is locally protected by a 13 A fuse or 16 A circuit breaker. This ensures that the connection cable is protected and cannot be overloaded. Guidance Note 1 states that the number of non-fused spurs is not to exceed the total number of socket-outlets and items of stationary equipment connected directly in the circuit, although there is no technical justification for this limitation.

Note that there are nine separate conductors in the socket-outlet where the unfused spur is connected. This can be difficult to accommodate in a single mounting box, particularly if the ring circuit is wired in 4 mm² cable.

Fused spurs are protected by the fuse in the connection unit and there is no limit on the number of socket-outlets on each. The number of fused spurs is unlimited.

The Wiring Regulations do not specify how many socket-outlets should be provided, only that consideration should be given to the length of flexible cord normally fitted to portable appliances and luminaires.

The current prevalence of IT equipment in commercial installations and both IT and audio–visual (AV) equipment in domestic installations requires a large number of socket-outlets to be installed, even though the total connected load is relatively low. For example, a typical home study of, say, eight square meters in area may have as many as 16 or 17 socket-outlets in use.

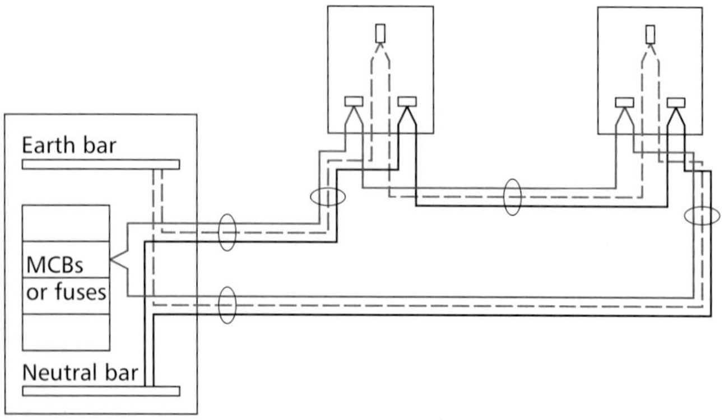

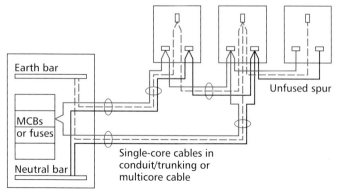

Figure 8.1 Basic ring final circuit (*Note*: cable colours will change from 2006, see section 8.4.1.4)

Figure 8.2 Basic ring final circuit with unfused spur (*Note*: cable colours will change from 2006, see section 8.4.1.4)

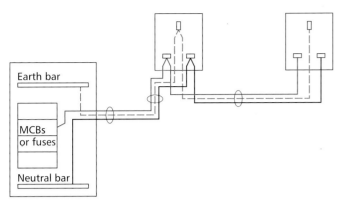

Figure 8.3 Basic radial final circuit; typical UK practice (*Note*: cable colours will change from 2006, see section 8.4.1.4)

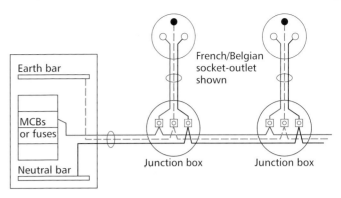

Figure 8.4 Basic radial final circuit; typical European practice (*Note*: cable colours will change from 2006, see section 8.4.1.4)

In a commercial or office installation the number of final sub-circuits is best determined by the layout of the desks or furniture. Where a three-phase final distribution board is installed the circuits should be balanced between the three phases. Selecting the number of circuits using the area loading (i.e. the watts per square metre figure) will normally result in a very low number of final circuits.

The physical route of the cables to and from the socket-outlets may be vertically from above or below. The Wiring Regulations dictate the routes that the cables must take in different types of installation.

8.3.3.2 Radial final circuits (UK practice)

Radial final circuits are used on lighting circuits and also for small power supplies in both domestic and commercial installations. These would typically be used to supply a single item of equipment such as an electric cooker or an underfloor distribution busbar. Multiple socket-outlets are usually connected as shown in Figure 8.3. The Wiring Regulations cover the cable sizes and allowable protective device ratings allowed for the various types of circuit and socket-outlet. As with ring circuits, the connections to the socket-outlets may be run from above or below.

8.3.3.3 Radial final circuits (European practice)

Radial final circuits are used universally throughout continental Europe, generally protected at 16 A and incorporating what are commonly known as 10/16 A socket-outlets because some earlier plugs were rated at 10 A rather than the 16 A rating of the socket-outlet. Several different types of socket-outlet are in use with the most common being the CEE 7/4 'Shuko' type, with the earth contacts on the side of the plug, used in Germany, Australia, the Netherlands, Sweden, Norway and Finland. France and Belgium use an incompatible socket-outlet with a male earthing pin, although there is a CEE 7/7 plug that works with both these socket-outlets. The Shuko plug can be inserted either way round, as can the Italian CEI 23-16 10 A and 16 A plugs, so line and neutral are connected at random in these systems.

Although European radial circuits can be wired as in the UK, with the cabling running from one socket-outlet to the next, it is more common to spur each socket off the cable run in a junction box as shown in Figure 8.4 In domestic installations these junction boxes are often mounted just below ceiling level and frequently have just a push-fit plastic cover.

8.3.3.4 Lighting circuit wiring methods

The methods used in wiring lighting circuits are often dependent upon the type of cabling used in the installation, particularly where two-way switching is required. The most common system in modern domestic and some commercial installations uses a ceiling rose with 'loop-in' terminals, connected as shown in Figure 8.5 The basic one-way switching circuit can be used with either single core or multicore cables, although two-way switching is wired quite differently, as shown in Figure 8.6, to avoid having to install loose connectors inside the switch box.

One point to note when using multicore cables is that the colour of the insulation may not be in accordance with BS 7671, which (up to April 2004) required the phase conductor of a single phase circuit to be red. This is

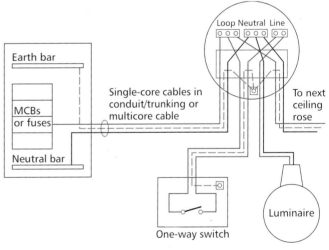

Figure 8.5 Basic lighting radial circuit (*Note*: cable colours will change from 2006, see section 8.4.1.4)

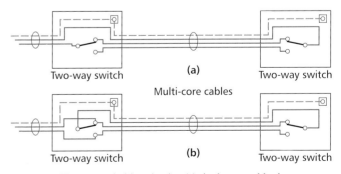

Figure 8.6 Two-way switching circuits; (a) single core cables in conduit/trunking, (b) multicore cables (*Note*: cable colours will change from 2006, see section 8.4.1.4)

particularly significant where 'flat twin and earth' type cables are used in domestic installations.

In the two-core cable connecting the ceiling rose to the switch, or the first switch in the case of two-way switching, both conductors are classified as live conductors. However when a standard cable is used one of the conductors will be black. This can be particularly confusing for amateur electricians. The accepted practice is to cover the black conductor with red sleeving but this is frequently not done. Some manufacturers produce two-core cables with two red conductors, but these are seldom seen.

Similarly, the three-core cable between the two-way switches will be insulated in the red, yellow and blue used for three-phase circuits. Technically these colours should only be used on a three-phase circuit but, at least, there is no live conductor with black insulation.

Note: the colours traditionally used in the UK are changing to EU harmonised colours, see section 8.4.1.

8.4   Power cables

8.4.1   Identification colours

The Wiring Regulations[1] define the identification colours for various type of conductors including those for direct current circuits. This section covers only those colours used in single phase and three-phase services. At the time of writing (May 2004) these colours are in the process of being changed.

8.4.1.1   UK situation (up to 2004)

For non-flexible cables the colours are:

—   phase of AC single phase circuit: red★

—   neutral of AC single or three-phase circuit: black

—   phase R of three-phase AC circuit: red

—   phase Y of three-phase AC circuit: yellow

—   phase B of three-phase AC circuit: blue

—   protective conductor ('earth'): green/yellow

—   functional earth: cream

For flexible cables and cords the colours are:

—   phase: brown (or black in 4- and 5-core cables)

—   neutral: blue

—   protective conductor ('earth'): green/yellow

8.4.1.2   Continental Europe situation (up to 2004)

For both non-flexible cables and flexible cables and cords the colours are:

★ In the Republic of Ireland red, yellow and blue are allowed for the final subcircuits. This has been observed in several installations in the UK where Irish contractors have carried out the installation.

—   phase of single phase circuit: black

—   phases of three-phase circuits: black, brown, black

—   neutral: blue

—   protective conductor ('earth'): green/yellow

8.4.1.3   Anomalies with the existing situation (up to 2004)

There are many difficulties and anomalies in the above systems.

In the UK, blue was used as both a phase and a neutral conductor, depending solely on whether the cable was flexible or not.

In continental Europe, the phase rotation is not immediately apparent, although cables can be phased out with some difficulty because one black conductor is next to the brown conductor and the other is next to the blue.

Probably the most confusing and possibly dangerous situation occurs when installing a continental European suspended luminaire in a UK installation because the fixed wiring has a black neutral conductor and the luminaire has a black live conductor.

8.4.1.4   Harmonised cable colours (after 2006)

A new system of harmonised cable colours has now been agreed. In the UK these colours were introduced on 1st April 2004 as an amendment to the Wiring Regulations[1]. However, they will not become compulsory until 1st April 2006. Installations commencing between these dates may use either of these colour systems, but not a combination of the two.

For both non-flexible cables and flexible cables and cords the colours are:

—   phase of single phase circuit: brown

—   neutral of AC single or three-phase circuit: blue

—   phase 1 of three-phase AC circuit (L1): brown

—   phase 2 of three-phase AC circuit (L2): black

—   phase 3 of three-phase AC circuit (L3): grey

—   protective conductor ('earth'): green/yellow

—   functional earth ('clean earth'): cream

The amendment to the Wiring Regulations of 1 April 2004 also gives details of how the new colour cables should be identified when making alterations or additions to installations that use the old colours.

8.4.2   Cable voltage rating

The voltage rating for cables is expressed as two values, for example 600/1000 V. The first figure is the maximum voltage allowable between any conductor and earth, and the second is the maximum voltage allowable between any two conductors, generally between the phases in a three-phase system.

The cables typically used in domestic installation have a voltage rating of 300/500 V and the larger power cables normally have a voltage rating of 600/1000 V.

It can thus be seen that where installations use the European standard voltage of 230/400 V there is no problem with either type of cable. However care must be taken where industrial installations use higher voltages, such as the 380/660 V systems proposed in Germany in the 1980s, because the phase-to-earth voltage may exceed the rating of some cable types.

8.4.3   Selection criteria

8.4.3.1   Electrical characteristics

The two main factors to be taken into account when determining the size (or, more correctly, the cross-sectional area (CSA)) of any cable are the maximum continuous current rating and the voltage drop in the circuit. The maximum current rating is determined by the maximum temperature rating of the insulation and depends on the way the cable is installed, e.g. whether it has air flowing freely around it or if it is enclosed in a conduit or duct. Other factors include the maximum circuit impedance for operation of short-circuit and earth fault protection devices and the ability of the cable to withstand the electromechanical stresses due to such faults.

The Wiring Regulations[1] and associated Guidance Notes[2,4–7] give details on how to calculate the required size of cable but most designers now use computer programs that take into account the various inter-related factors.

While both the maximum current rating and the voltage drop must be considered for all circuits it will generally be found that one or other is the major factor in determining the required cable size.

In general the cable CSA on high current circuits such as the connection between a transformer and the main switchboard, or on a short sub-main circuit to a large motor control centre, will be determined by the maximum current rating, whereas long sub-mains and long final circuits will be sized to keep the voltage drop within acceptable limits.

8.4.3.2   Mechanical characteristics

BS 7671[1] includes a comprehensive list of all the external influences to be taken into consideration when designing the electrical installation. The main ones relevant to cable selection are temperature and protection against mechanical damage.

The current rating of a cable is determined by the maximum temperature rating of the insulation. The current ratings included in BS 7671 and manufacturers' data are quoted under defined installation methods and ambient temperatures. When installed under these defined conditions the heat generated by the current flowing in the cable will be dissipated at a rate such that the allowable maximum temperature is not exceeded.

However, should the cable be installed in a manner which prevents this heat from dissipating quickly enough, e.g. if the cable is installed within thermal insulation or if the ambient temperature is very high, then the cable will become too hot and the insulation will start to degrade. This situation will also arise if the cable is installed where heat from other building services can be transferred to the cable, as may occur where cables are run adjacent to hot water pipework. It will also occur where many cables are installed close to one another.

The necessary degree of protection against mechanical damage will vary considerably according to the nature of the building use and the likelihood of the installation being subject to physical impacts. Obviously the cables in a warehouse with mechanical handling equipment will need greater protection than those in a domestic attic.

Adequate mechanical protection can be afforded by the correct choice of cable type or containment system and these are covered in more detail in the appropriate sections.

8.4.3.3   Environmental conditions

Underground installation

Cables can be installed underground either by burying them directly or pulling them into ducts. Cables in trenches are frequently laid on a bed of sand so there are no sharp stones that might damage the cable sheath and allow water to penetrate. Duct systems will generally have manholes to allow access and these must not be spaced too far apart in order to avoid excessive strain on the cable when pulling it into the duct.

The correct choice of a cable for direct burial depends on the soil conditions, e.g. whether there are any corrosive elements, and the outer sheath of the cable. Most sheathed single core and multicore power cables are suitable for installation directly in the ground, although the manufacturers should always be consulted when the soil conditions may be aggressive.

In the UK, it is normal to use armoured cables for underground installations although in Europe non-armoured cables are the norm.

Underwater installation

Underwater installation covers two distinctly separate areas:

—   Where the cable supplies a water feature or submersible pump it is obviously intended to be continuously immersed in water and the cable type will be chosen accordingly, often by the equipment supplier.

—   Where a cable in a trench or duct is likely to be immersed for long periods, the manufacturer should be consulted to ensure that the correct cable type is used.

8.4.3.4 Life expectancy

There is no hard and fast definition of the 'life' of a cable. In some cases, the periodic inspection and testing will indicate when the insulation has started to fail and this will indicate that the cable needs to be replaced. In other cases the cable may give adequate readings during tests but the insulation may have become brittle and could fall off the conductor when alterations or additions are made to the installation.

Earlier types of cables, such as those with vulcanised rubber insulation that were installed in the 1950s have reached the end of their working life. However the life expectancy of modern PVC and XLPE cables is still largely unknown. Many such cables have been in service for over thirty years and show no signs of deterioration. However, a notable exception is the first low smoke and fume (LSF) cables that were produced in the early 1980s. Manufacturers were still experimenting with different insulation compounds and these cables were not as good as their present day equivalents.

All the above assumes that the cables are installed properly and not subject to any adverse conditions. Probably the major cause of cable insulation failure is localised heating caused by poor joints or loose terminations, or where the cables have been installed in contact with incompatible materials.

An example of this latter problem is where polystyrene thermal insulation was installed in domestic lofts in contact with PVC/PVC 'flat twin and earth' wiring. It was subsequently discovered that the polystyrene caused the plasticiser to migrate from the PVC sheathing of the cables and made them become brittle.

8.4.4 Cable conductors

8.4.4.1 Materials

The most common cable conductor material used in building installations is copper. In some types of cables the copper may be tinned to prevent the copper from reacting with the insulation material. Aluminium is often used for power cables, particularly by the electricity companies. These require different termination and jointing techniques. They are not commonly used in building installations and therefore are not considered in this Guide.

Cables for specialist applications such as trace heating or underfloor warming may use materials such as copper-nickel or nickel-chromium alloys.

8.4.4.2 Construction

BS 6360[8] defines four classes of cable conductor:

— class 1 has solid conductors

— class 2 has stranded conductors

— classes 5 and 6 are flexible cables and cords, with class 6 being more flexible than class 5.

The classes apply to both single core and multicore cables.

The British Standards for the various cable types cover which classes of conductor are allowable for the different cross sectional areas. The standards covering most of the cables used for fixed wiring can have either solid or stranded conductors up to 10 mm² CSA. However, in practice it will be found that these cables generally have solid conductors only for cables up to 1.5 mm² or 2.5 mm² and stranded conductors for the larger sizes.

Flexible conductors are commonly found in small sizes in multicore cable connections to luminaires and fixed equipment but are also used in large single core cables for the final connections to vibrating equipment such as generator sets.

8.4.5 Insulation

Insulation materials have evolved over the years and will probably continue to do so in the future. Many types of insulation materials are available particularly for specialist applications, but only the more common ones will be covered here.

8.4.5.1 Polyvinyl chloride (PVC)

PVC insulation was used extensively from the 1960s up until the early 1990s, when concerns about the toxicity of combustion products lead to it going out of common use. It is probably still the most commonly found insulation material and was suitable for both fixed and flexible installations. It is a clean, easy to handle material with good electrical characteristics and a reasonable resistance to a range of possible contaminants such as water, oils and chemicals. It is inherently flame retardant and is suitable for a conductor maximum continuous operating temperature of 70 °C and a maximum short-circuit conductor temperature of 160 °C (140 °C for conductors larger than 300 mm².

Most UK and European manufacturers still list cables with PVC insulation.

8.4.5.2 Thermosetting materials (XLPE and EPR)

The type of thermosetting insulation most commonly referred to is cross-linked polyethylene (XLPE) although all the British Standards covering this type of cable permit either XLPE or ethylene propylene rubber (EPR) as an alternative insulation material.

XLPE is superior to PVC in that it is able to operate at higher temperatures with a sustained conductor temperature of 90 °C as opposed to the 70 °C of PVC. The maximum short-circuit conductor temperature is 250 °C. It also has better insulating properties that achieve the same voltage ratings with thinner insulation.

8.4.5.3 Mineral insulation (MI)

Mineral insulated cables have the mineral insulation tightly packed between a solid bare copper conductor (or conductors) and a tubular copper sheath. Magnesium oxide has commonly been used as the insulating medium but the current version of BS EN 60702-1[9] specifies only the insulation properties, not the material to be used.

The main advantages of this type of construction are that it will not burn or support combustion or produce toxic or corrosive fumes in a fire. MI cables are able to operate at much higher temperatures than either PVC or XLPE cables. In principle, the cables themselves will still operate at temperatures approaching the melting point of copper, over 1000 °C, making them ideal for cables that have to survive under fire conditions. In reality the maximum operating temperature is limited by the end seals that have to be use to seal the cables against ingress of moisture.

Because the cable itself is unaffected by operation at extremely high conductor temperatures it is necessary to used a different method of determining the maximum current rating. BS 7671[1] defines two operating conditions for MI cables. Where the outer sheath is exposed to touch, the maximum sheath temperature is limited to 70 °C and where the sheath is not exposed to touch the maximum sheath temperature is allowed to rise to 105 °C.

Note that MI cables always have solid conductors and that this requires different jointing and termination techniques. It also means that the larger sizes of cables are impossible to bend by hand.

8.4.5.4   LSF and LS0H insulation and sheaths

There are various abbreviations for cables that give off very low emissions of smoke and fumes when subject to fire conditions, although none of them is officially defined in a British Standard. LSF ('low smoke and fume') and LS0H ('low smoke, zero halogen') are probably the most common. The insulation materials used in these cables are defined in the relevant standards but are not generally referred to when specifying the cables.

8.4.5.5   'Fire-resistant' cables

Many of the systems in a large building are required to continue to operate in the event of a fire. Typically these will include the fire detection and alarm system itself, emergency lighting, and the power supplies to firefighting equipment and lifts.

The combinations of manufacturing standards and performance tests are complicated and care must be taken when specifying cable types. Many designers simply specify cables with the highest available ratings, inevitably MI, although in many cases this may result in unnecessary expense for the client.

The following outlines the recent history of fire-resistant cables in the UK. The references to the relevant standards are necessarily short and reference should be made to the complete standard where necessary.

BS 6387[10] was issued in 1983 and set out the performance standards and tests for cables under fire conditions but was not a cable manufacturing standard.

The standard covers three aspects of cable performance:

—   resistance to fire alone

—   resistance to fire with water

—   resistance to fire with mechanical shock.

It details all the relevant tests for assessing the cable in all three categories but does not specify that the three tests are to be carried out on the same sample of cable. This leads to much public argument between the manufacturers of MI cables, which could pass all the tests using the same cable sample, and other manufacturers whose cables could pass the tests individually on separate cable samples.

The first British Standard for fire-resistant cables, BS 7629[11], was published in 1993 and referred to the cables as having 'limited circuit integrity' under fire conditions. This was changed to 'fire-resistant' in the 1997 version. The standard covers multicore cables with up to 4 mm^2 conductors as typically used in fire alarm and emergency lighting installations. Cables such as Pirelli FP200 and Tyco Pyro S are manufactured to BS 7629-1.

BS 7846[12] covers fire-resistant armoured power cables with conductor sizes up to 400 mm^2. At the time of writing (May 2004) there are few cables certificated by the Loss Prevention Certification Board (LPCB) as meeting this standard.

Both BS 7629 and BS 7846 require the cables to have LSF/LF0H characteristics.

The subject of fire resistance was further complicated in 2000 by the adoption of a new European standard, BS EN 50200[13]. This defines fire resistance tests for small cables but crucially does not include a water test. This was addressed in 2003 by BS 8434: Part 1: *BS EN 50200 with addition of water spray*[14].

The current edition of the fire alarm standard also specifies additional water spray tests for cables meeting the relevant BS EN 50200 classifications. Interestingly, MI cables manufactured to BS EN 60702[15] are included as acceptable without further comment.

In many installations the type of cable will be subject to approval by the Fire or Building Control Officer, and it is not uncommon for them to adopt a cautious approach and insist on MI cables.

8.4.6   Multicore cables

The term 'multicore' is used to describe a cable where two or more separately insulated conductors are contained within a single outer sheath.

Cables for electrical distribution in the UK normally have between two and four conductors, European practice frequently uses five-core cables, see section 8.4.6.3. Cables for signalling and control systems may have many more conductors.

The range of multicore cables is vast, extending from small two-core cables used on lighting installations up to power cables with conductors of several hundred square millimetres cross sectional area.

Probably the main advantage of multicore cables is that they can be installed directly onto a wall or cable tray and frequently do not need any other form of protection.

8.4.6.1 Naming conventions for multicore cables

Both single core and multicore cables are frequently referred to by a string of abbreviations describing the construction of the cable, starting with the conductor insulation and moving outwards. Using this system a single core PVC cable with a PVC sheath would be referred to as 'PVC/PVC' and a multicore cable with XLPE insulation, steel wire armouring and a PVC sheath would be referred to as 'XLPE/SWA/PVC'.

Codes used in the UK

In the UK, the Cable Manufacturers' Association developed a series of number codes for different cable types and these are commonly used by electrical contractors. Examples of some of the more commonly used codes are given in Table 8.1.

Codes used in the EU

Cables that comply with the European harmonized standard are marked with the symbol '<HAR>'.

The designation system uses a series of up to ten letter/number codes to indicate the construction of the cable. These appear in a fixed order, see below. The code relating to the type of conductor is preceded by a hyphen and followed by a space.

The system can appear somewhat confusing in practice because unused designation codes are simply omitted. For example, a 5-core, 2.5 mm² 300/500-volt oil resisting PVC sheathed, screened cable to BS 6004 is designated H05VVC4V5-K 5G2.5, whereas a single core stranded conduit wire would be designated as H07V-R.

The codes, in order of appearance, are as follows:

(1) *Basic standards*

H: harmonised standards
A: authorised national standards

(2) *Rated voltage*

03: 300/300 volts
05: 300/500 volts
07: 450/750 volts

(3) *Insulation material*

B: ethylene propylene rubber (EPR)
E: polyethylene, low density (LDPE)
E2: polyethylene, high density (HDPE)
E4: polytetrafluoroethylene (PTFE)
E6: ethylene tetrafluoroethylene (ETFE)
E7: polypropylene (PP)
G: ethylene vinyl acetate (EVA)
J: glass fibre braid (GFB)
N: polychloroprene (PCP)
N4: chlorosulphonated polyethylene (CSP)
R: natural rubber
S: silicone
T: textile braid
V: polyvinylchloride (PVC)
V2: heat resistant PVC
V3: low temperature PVC
V4: cross-linked PVC
V5: oil resistant PVC
X: cross-linked polyethylene (XLPE)

(4) *Structural elements*

C4: overall copper braid screen
C5: cores individual copper braid screen
C7: lapped copper (wire, tape or strip) screen

(5) *Sheath material*

Codes generally as for (3), insulation material

(6) *Special constructions and shapes*

D3: central strainer (textile or metallic)
D5: central filler (not load bearing)
H: flat construction with divisible cores
H2: flat construction, non-divisible cores

Table 8.1 Common UK cable references

Reference	Description
6242Y	PVC/PVC 'flat twin and earth' domestic wiring cables to BS 6004[16]
6242B	XLPE/LSF 'flat twin and earth' domestic wiring cables to BS 7211[17]
6243Y	PVC/PVC 'flat triple and earth' domestic wiring cables to BS 6004[16]
6243B	XLPE/LSF 'flat triple and earth' domestic wiring cables to BS 7211[17]
6491X	PVC insulated single core cable to BS 6004[16]; the normal single core PVC cable for installation in conduit
6491B	LSF insulated single core cable to BS 7211[17]; as 6491X but with LSF insulation, for installation in conduit
6181Y	PVC/PVC single core unarmoured cable to BS 6004[16]; cross sectional area up to 35 mm²
6181Y	PVC/PVC single core unarmoured cable with cross sectional area 50 mm² and above; see note 1
6181XY	XLPE/PVC single core unarmoured cable to BS 7889[18]; see note 2
6181B	XLPE/LSF single core unarmoured cable to BS 7211[17], cross sectional area up to 35 mm²
6941X	XLPE/AWA/PVC single core armoured cables to BS 5467[19]
6941B	XLPE/AWA/LSF single core armoured cables to BS 6724[20]
694*X	XLPE/SWA/PVC multicore armoured cables to BS 5467[19]
694*B	XLPE/SWA/LSF multicore armoured cables to BS 6724[20]

* indicates the number of cores in a multicore cable and can be 2, 3 or 4

Notes:

(1) PVC/PVC unarmoured cables were originally covered by BS 6346 but unarmoured cables were removed in the 1997 edition.

(2) XLPE/PVC cables were originally covered by BS 5467 but unarmoured cables were removed in the 1997 edition. Some cable suppliers still indicate that these cables are manufactured to BS 5467: 1989.

H5: two or more cores twisted together, non-sheathed

(7)   *Type of conductor*

The classes of conductor are as defined in BS 6360[8]:

U: solid (class 1)
R: stranded (class 2)
K: flexible for fixed installations (class 5)
F: flexible for moveable installations (class 5)
H: highly flexible for moveable installations (class 6)

(8)   *Number of cores*

(9)   *Protective conductor*

X: without protective core
G: with protective core

(10) *Nominal cross section of conductors*

Value stated in mm².

8.4.6.2   Cables for domestic installations

The most common multicore cables used in domestic installations in the UK are the 'flat twin and earth' types that incorporate two or three insulated conductors with or without an uninsulated CPC (earth) conductor. Cables are available either with PVC insulation and sheath or in an LSF version with XLPE insulation and an LS0H sheath.

Conductors up to 2.5 mm² CSA are solid; larger sizes are stranded, as shown in Figure 8.7.

These cables are manufactured to BS 6004[16] for PVC/PVC construction and BS 7211[17] for LSF/LS0H construction, with conductor sizes from 1.0 mm² to 16 mm². The most commonly used sizes are 1.0 and 1.5 mm² for lighting circuits and 2.5 and 4 mm² for small power circuits. Larger sizes are used for higher power circuits such as cookers and electric showers.

8.4.6.3   Power distribution cables

These include the mains and sub-mains inside a building, supplies to plant and the buried supply cables to the building. These cables are available with several types of insulation and sheathing materials in various combinations.

In the UK most cables will be armoured with a layer of steel wires that provide both mechanical protection and are also used as the CPC or 'earth'. These installations will use two-core cables on single phase circuits and four-core cables, with the steel wire armouring providing the CPC. The neutral conductor in a four-core cable will normally

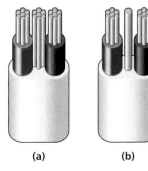

Figure 8.7 Multicore cable; domestic 'flat twin and earth' (*Note*: cable colours will change from 2006, see section 8.4.1.4)

(a)   (b)

be the same size as the phase conductors although cables with a reduced size neutral are available for use on systems where the three-phase load is well balanced across the phases. Such cable are sometimes referred to as 3½ core because the neutral conductor CSA is about half the phase conductor CSA. The armouring is usually referred to as SWA although this is variously taken as meaning 'steel wire armour' or 'single wire armour' by different manufacturers.

In continental Europe power cables generally include a dedicated conductor for the CPC so a typical three-phase, neutral and earth cable will have five separate cores. In the larger cable sizes the CPC is normally half the size of the phase conductor. The neutral may be full or half size depending on the application.

For many years cables having PVC insulation and a PVC sheath were used practically universally, with XLPE insulation being used where higher current ratings were required. Nowadays PVC is used much less frequently and most installations use cables with XLPE insulation and an LSF or LS0H sheath.

Typical armoured power cables are shown in Figure 8.8.

Figure 8.8 Multicore armoured power cables (*Note*: cable colours will change from 2006, see section 8.4.1.4)

(a)   (b)

8.4.6.4   Flexible cables and cords

These cables are often used for the final connections to portable or fixed equipment in domestic, commercial and industrial installations. Multicore cables are available with between two and five cores according to the manufacturing standard.

In the UK there are two standards for flexible cables and cords:

—   BS 6500[18] covers cables for office and domestic environments and lists several cable types suitable for light or ordinary duty. These cables generally have PVC or EPR insulation and PVC or CSP sheaths.

—   BS 7919[19] covers cables for industrial use and includes a wide range of insulation and sheath materials, including heat resisting, oil resisting, flame retardant (HOFR) and LS0H materials.

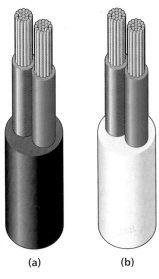

Figure 8.9 Flexible cords (*Note*: cable colours will change from 2006, see section 8.4.1.4)

(a)   (b)

Some types of flexible cords are available in high temperature variants for connection to heaters or installation in high ambient temperatures. Typical flexible cords are shown in Figure 8.9.

8.4.7   Single core cables

The most common area where single core cables are used is in final sub-circuits supplying small power and lighting circuits. The cables are installed in an enclosed system of conduits, possibly with trunking used where many cables run together. The cables used on final sub-circuits are normally unsheathed.

Single core cables are also frequently used for the connections between transformers and the main switchgear. The very high currents involved require large cables, possibly with several in parallel, and single core cables are easier to install than similarly sized multicore cables. Sheathed cables will be used in most cases.

The insulation materials used for single core cables are similar to those described above for multicore cables, however single core cables are available both with or without a covering sheath. Sheathed cables are normally used where the cables are not enclosed inside a containment system, Figure 8.10.

Large single core cables are available with wire armouring for use where additional mechanical protection is required. The armouring has to be a non-ferrous metal to avoid heating by induced currents and aluminium is normally used.

Figure 8.10 Single core cables; (a) sheathed, (b) unsheathed (*Note*: cable colours will change from 2006, see section 8.4.1.4)

(a)   (b)

8.4.8   Other cable types

8.4.8.1   Metal clad cables

This type of cable is basically a group of single core cables pre-assembled into a flexible steel conduit as shown in Figure 8.11. Metal clad (MC) cable has been used for many years in North America but has only recently been used in the UK. At present there is no applicable British Standard although several MC cables have been certified by BASEC.

Termination is very simple with the external sheath being cut to length and clamped into a proprietary gland. A small plastic bush is installed over the cut end of the metal cladding to prevent damage to the conductor insulation.

MC cable is available in three, four and five core versions in sizes up to 35 mm².

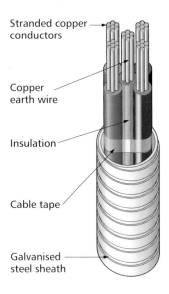

Stranded copper conductors

Copper earth wire

Insulation

Cable tape

Galvanised steel sheath

Figure 8.11 Five core metal clad (MC) cable (*Note*: cable colours will change from 2006, see section 8.4.1.4)

8.5   Cable installation, termination and jointing

8.5.1   Bending radius

Cables are very rugged and are not likely to be damaged during normal installation provided reasonable care is taken. However, the bending radius is very important. Cables must not be bent below a certain radius or damage will occur to the insulation and, possibly, to the bedding or sheathing in multicore cables.

Manufacturers quote the minimum bending radius as a multiple of the overall diameter of the cable, normally about eight times the diameter. With large multicore armoured cables it is difficult to bend the cables to the minimum bending radius without a former and, possibly, mechanical assistance. However, with smaller cables, up to about 15 or 20 mm overall diameter, it is quite easy to bend the cable to a smaller radius by hand and it may suffer damage as a result.

The necessary bending radius is frequently overlooked by designers, particularly where large armoured cables are terminated into a main switchboard. The cable needs to be straight where it enters the gland so that the gland can anchor all the armouring wires and make a good electrical

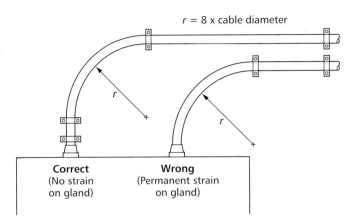

Figure 8.12 Correct and incorrect cable entry to switchboard

connection with the earth tag. The cable should not start to bend where it enters the gland or the gland will be permanently under strain. It is even possible that the gland plate on the switchboard will be deformed. This is shown in Figure 8.12.

The larger sizes of mineral insulated (MI) cables are impossible to bend by hand and need to be bent using a former or bending tool similar to those used for bending copper tubing or conduit. Most manufacturers recommend a minimum bending radius of six times the overall diameter, although the current British Standard requires cables to be tested only to a radius of about 12 times the overall diameter.

8.5.2   Terminations

The word 'termination' covers both the connection of the conductors to the terminals of the equipment and, in the case of multicore cables, the mechanical connection of the cable sheath and armouring.

The type of conductor termination will often be determined by what the manufacturers include as standard in their equipment. Most low current equipment and wiring accessories will incorporate simple pinch screw terminals and these are commonly used with conductors of up to 6 mm^2 CSA.

For larger conductors up to about 16 or 25 mm^2, screw clamp terminals are common. Above these sizes compression lugs are normally used. These are available in a wide variety of configurations to suit different applications, see Figure 8.13. Annex D of BS EN 60947-1[20] includes a description of the many types of terminals available.

Compression lugs should always be installed using the correct size dies and crimping tool. Tools for small size

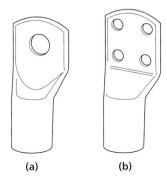

(a)   (b)

Figure 8.13 Typical compression lugs

lugs may be hand operated but those for large lugs are normally hydraulic or battery operated. The crimping tool should incorporate some form of non-return system to ensure that it cannot be removed before the crimping action is complete.

8.5.2.1   Cable glands

Where multicore cables are used it is normally necessary to install a cable gland where the cable enters the equipment enclosure. The cable gland provides a firm mechanical anchor for the cable sheath to prevent movement and possible damage to the conductor terminations and in the case of armoured cables it provides a connection to the armour wires so they can be used as the CPC.

Glands for flexible cords are generally made of plastic or nylon. Glands for armoured power cables, see Figure 8.14, are normally brass although other materials are sometimes used where metallic compatibility may be a problem. To ensure a good electrical connection to the gland an earth tag or an integral earth stud is provided.

The manufacturers indicate the IP rating of the assembled gland but this does imply that the gland has been correctly fitted to the cable. Additional protection can be obtained by fitting shrouds over the gland assembly and this is recommended for outdoor installations.

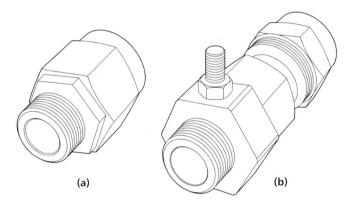

(a)   (b)

Figure 8.14 Glands for armoured cable; (a) plain, (b) with integral earth stud

8.5.3   Joints

The general consensus among designers is that any joint is a possible source of failure and that joints should be avoided unless absolutely necessary. This opinion is not shared by the electricity supply companies which have many thousands of underground joints in their networks that rarely cause problems.

The Wiring Regulations[1] do not define the term 'joint' although they do include specific requirements for where joints can be located. For the purposes of this section, a joint can be defined as a connection between two or more cables, as distinct from the connection of a cable into an item of equipment.

Joints will normally only be necessary in one of two circumstances:

—   where the required length of cable is not available in a single length from the manufacturer

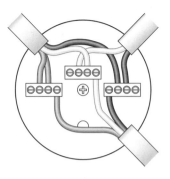

Figure 8.15 Junction box for domestic wiring (cover removed) (*Note*: cable colours will change from 2006, see section 8.4.1.4)

— where it is necessary to connect tee-off cables onto a main supply cable, as typically found in domestic supply systems.

8.5.3.1 Joints within buildings

Where either type of joint is required inside a building the conventional solution is to install a dedicated box with fixed terminals and connect the cables to the terminals using the termination method appropriate to the size of cable.

For multicore cables in a domestic installation this will involve a commercially available junction box as shown in Figure 8.15. For large cables such as may be used on a main riser system it may be necessary to install a purpose-made connection box with insulated terminals. In both cases the enclosure should be accessible so the connections can be inspected and checked for tightness, overheating or other problems.

8.5.3.2 Underground joints

These joints are sealed and inspection is not possible once the joint is completed.

The conductors are connected using either bolted or crimped connectors. Different types of connectors are available depending on whether the joint is a straight-through or tee-off. Connectors for the latter case are frequently split so there is no need to cut the main cable.

The finished joint needs to be encapsulated to provide mechanical protection and prevent the ingress of moisture. The most common method is to encapsulate the joint in a resin compound poured into a plastic mould as shown in Figure 8.16. Heat shrink tubing is an alternative form of encapsulation and is particularly suitable for straight-through joints.

8.5.4 Mineral insulated cables

MI cables require different jointing and termination techniques. The cables ends must be sealed to prevent the ingress of moisture and this is done using proprietary pot

seals supplied by the cable manufacturer. These are required for both single core and multicore cables.

As indicated earlier MI cables always have solid conductors. Cables up to about 10 mm² can be terminated with the same compression lugs as used on stranded cables but this is not possible for the larger sizes of conductor and a mechanical lug must be used.

8.6 High voltage installations

8.6.1 General

There are many types of high voltage cables capable of operating at voltages from a few kilovolts up to several hundred kilovolts. This section deals only with those likely to be encountered in a typical building installation.

The installation of the actual cable is very similar to the installation of an LV cable. The usual precautions about bending radii and secure anchoring of single core cables are equally applicable. The main difference is in the termination and jointing of HV cables.

With HV cables there must be a gradual change of potential across the insulation at the exposed end of the cable. This is known as 'stress relief' and prevents sudden changes in potential that could lead to corona or other discharge effects.

8.6.2 Thermosetting plastic cables

These are available as single core or three-core with copper or aluminium conductors. Insulation is typically XLPE or EPR. The only significant differences from the LV equivalents are the thicker insulation and, on some cables, the addition of a metallic or semi-conducting screen around the conductor. Installation methods are the same as for LV cables.

Stress relief is usually provided by a semi-conducting sleeve that is slid over the exposed insulation. The sleeves and any filling compounds and other accessories are available as complete kits from the manufacturer. The actual sleeve may be slip-on, heat shrink or cold shrink. Installation is very simple with almost no specialist training required. Indoor terminations may be a plain tube; outdoor terminations usually have circular 'sheds' to

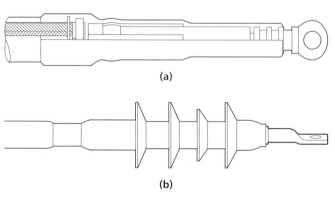

(a)

(b)

Figure 8.17 Typical 11 kV terminations; (a) indoor, (b) outdoor

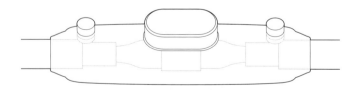

Figure 8.16 Underground joint

provide a greater creepage path. Typical terminations are shown in Figure 8.17.

Cables are available in both armoured and non-armoured versions. Where present the armouring is terminated in a metal gland exactly as for an LV cable.

8.6.3 Paper insulated cables (PILC)

These types of cable were used extensively by the electricity supply companies in their underground distribution networks. The insulation consists of layers of oiled paper wrapped around the conductor and the whole cable is contained in a continuous tubular lead sheath together with various fillers and bedding. With some types of cable precautions had to be taken to prevent the oil flowing along the cable on inclined runs. The mass impregnated, non-draining (MIND) cable was developed to overcome this problem.

The termination of this type of cable has traditionally been a highly skilled task. The outer lead sheath is plumbed onto a gland on the termination box in a similar way to a lead water pipe. Stress relief is applied to the conductors by wrapping them in layers of special tape, sometimes impregnated with carbon black. The correct application of these tapes is critical to obtain the correct performance. After completion the termination box may be filled with an insulating compound.

Manufacturers now produce termination kits using modern heatshrink materials although the installation of these on a PILC cable is still much more complicated than on an XLPE cable.

8.6.4 Sizing of HV cables

In a typical building installation the LV cables will be sized on either voltage drop or maximum current rating. This is hardly ever the case for HV cables.

At a supply voltage of 11 kV the full load current on a 1 MV·A transformer is about 52 A. The smallest size 11 kV cable has a 50 mm² CSA and a current rating of over 200 A. At first sight it would appear that this cable would be more than adequate to supply the transformer. However, the current that will flow in the cables in the event of a short circuit fault must be taken into account. The fault level in most large UK cities is quoted as 250 MV·A, or 13.1 kA at 11 kV and this is the current that would flow in the cable in the event of a fault.

Normally the fault current will be cleared by the protective device immediately upstream from the fault, but on HV networks it is common practice to consider the situation where this device fails to operate correctly so the fault current flows until the next upstream device operates. In a time graded system with 300–400 ms grading margin between adjacent devices this may mean that the fault current flows for nearly one second. The cable must be able to carry the current for this time without damage and it is customary to select the cable based on the one second short circuit rating.

A 50 mm² cable has a one second short circuit rating of only about 7 kA. The smallest cable with a one second short circuit rating above 13.1 kA is 95 mm² and this is generally accepted as the smallest size that can safely be used on an 11 kV 250 MV·A network.

8.7 Containment systems

8.7.1 Trunking

8.7.1.1 General

Trunking is available in many different configurations. The most common format is a rectangular tube with a removable lid, as shown in Figure 8.19. This type of trunking can be installed horizontally or vertically, mounted directly on walls or ceilings or suspended on various types of hangers. When wall mounted it is necessary to install restraints to prevent the cables falling out until the lid is installed. Other types of trunking include systems for burying in concrete floors and skirting and dado trunking incorporating sockets for power and communications circuits.

This section concentrates on the rectangular type of trunking that is typically installed from a final distribution board.

8.7.1.2 Steel trunking

Conventional steel trunking is manufactured from mild steel and is available in a range of sizes from 38 mm by 38 mm up to 300 mm by 300 mm. The section can be square or rectangular with various width-to-depth ratios. The largest sizes are rarely encountered because few installations require the number of conductors that these sizes can accommodate. The largest sizes likely to be encountered will be no larger than 100 mm by 225 mm, probably divided internally into two or three separate compartments.

BS 4678-1[21] defines the thickness of the metal to be used and includes three classes of protection against corrosion. Classes 1 and 2 require the steel to have an electroplated zinc coating and Class 3 is hot-dip galvanised, generally specified for external installations, although it should be remembered that very few trunking systems can be described as fully weatherproof and the BS makes no reference to IP ratings.

Manufacturers offer a complete range of fittings including bends, tees and reducers but installation still requires much cutting and drilling on site.

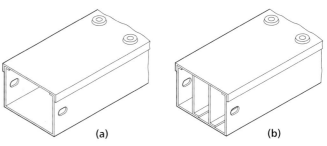

Figure 8.18 Trunking; (a) single compartment, (b) multi-compartment

Stainless steel trunking is available for specialist installations but is very expensive, in terms of both initial purchase and installation.

8.7.1.3   Plastic trunking

Plastic trunking (or, more correctly, 'cable trunking'), made of insulating material, is manufactured in a range of sizes from 12.5 mm to 150 mm as defined in BS 4678-4[21]. Manufacturers offer 'mini' trunking in a range of sizes from 16 mm by 16 mm up to 50 mm by 50 mm, and some offer larger sizes up to 100 mm by 75 mm.

BS 4678-4 includes classifications for various aspects of the trunking, including the material itself and its resistance to flame propagation. Unlike the standard for steel trunking it also includes categories for the ingress of solid objects and water, so the trunking can have a definite IP rating.

The smaller sizes of plastic trunking are primarily intended for surface installation of final subcircuits and many manufacturers produce adapters so that the trunking can connect directly to accessory mounting boxes. This type of installation would hardly ever be specified for a new installation but can be very practical for a small office refurbishment or additions to an existing installation.

8.7.1.4   Multi-compartment trunking

Both steel and plastic trunking are available in multi-compartment versions. Some manufacturers offer barriers to divide single compartment trunking into two or more separate compartments.

This type of trunking enables circuits that require segregation from each other to be installed in a single trunking system rather than having to install separate trunking systems for the different circuits. This type of installation can work well with simple trunking systems but can result in difficulties where complex layout are involved.

The main problem is that segregation must be maintained throughout the trunking system, so that at flat tee junctions the cables from one compartment need to pass through the other compartment. Installation contractors may prefer to install separate systems of single compartment trunking rather than a complicated system of multi-compartment trunking.

8.7.2   Conduits

8.7.2.1   General

In most installations using single core cables the final connections to items such as socket-outlets and switches will be installed in conduit. In small installations the conduits may run all the way from the distribution board; in larger installations conduits will connect back into a trunking system.

Conduits may be surface mounted, suspended on hangers or buried in walls and floors according to the requirements of the installation. Conduits are manufactured from both steel and plastic and may be either rigid, flexible or pliable, meaning that it can be bent by hand.

The current British and European Standard for both steel and plastic conduits, BS EN 61386[22] (Parts 1, 21, 22, 23), lists a range of twelve sizes ranging from 6 mm to 75 mm diameter for both rigid and pliable conduits. The old, but still current, British Standard for rigid conduit, BS 4568: Part 1[23], covered only 16, 20, 25 and 32 mm sizes. In the UK, only 16 and 20 mm conduits are commonly used with most installers preferring to use trunking where greater capacity is required.

8.7.2.2   Steel rigid conduit

Steel conduit is available in a heavy gauge which is screwed into the fittings and a light gauge which uses slip joints.

Screwed conduit has been used in the UK for many years and produces a very strong system with good resistance to mechanical damage although the cost of installation is high. As the system is all tightly screwed together the electrical continuity is very good and the conduit system can be used as the CPC (earth) conductor. The most common finishes in the UK are black enamel for internal installation and galvanised for external use.

Light gauge conduit systems are much quicker to install because no threading is required. However care must be taken to anchor the conduits adequately where it is to be buried in poured concrete as there is nothing to prevent the joints from separating. The slip joints do not guarantee electrical continuity and this type of conduit cannot be used as a CPC.

Some manufacturers produce a slip joint conduit system where the accessories and couplings have a spring steel insert that is designed both to prevent joints coming apart and to provide electrical continuity.

8.7.2.3   Plastic rigid conduit

Plastic conduit is available in both circular and oval cross sections, the latter frequently used for final connection to accessory boxes in narrow partitions and plaster skims.

Joints are slip fit but may be cemented using a liquid solvent where the conduit is to be buried in concrete.

8.7.2.4   Flexible conduit for final connections

Flexible conduit is frequently used for the final connections to moving or vibrating equipment such as door closers and electric motors. Flexible conduits may be of metal or plastic or a combination of several materials.

Products vary greatly between manufacturers and those from one manufacturer may be incompatible with those from another, or even with a different product range from the same manufacturer. Therefore it is important that the conduit and termination glands are sourced correctly.

Flexible conduits are frequently referred to as 'kopex', after the name of a UK manufacturer.

8.7.2.5 Flexible conduit for fixed wiring

Flexible conduit is commonly used for fixed wiring in Europe and the Middle East, where corrugated plastic conduit is embedded in walls and floors. In North America flexible steel conduit is used within studding walls. Neither of these practices is common in the UK.

8.7.2.6 Rewirable conduit systems

In the past many designers specified that conduit systems had to be rewirable. The reasons for this are unclear but it probably originates from the time when the life expectancy of the cable insulation was only about ten or 20 years.

Nowadays there is no need to specify rewirable systems. The electrical installations in most offices are often modified or completely changed well within the life expectancy of the cables and the original conduits will rarely be in the correct locations.

8.7.2.7 Open conduit systems

In open conduit systems the conduit is used solely as a support system for multicore cables. Conduits terminate about 50–100 mm from equipment or junction boxes and the cable enters the enclosure through a plastic gland. This type of wiring is extremely common in continental Europe but is not found in the UK.

8.8 Cable support systems

The current British and European Standard (BS EN 61537[24]) covers traditional cable tray, basket tray and cable ladders.

These types of systems are most appropriate where large numbers of cables are distributed in the horizontal plane. They can be installed vertically but where large cables are installed there may not be any advantage over cleating the cables directly to a channel support system.

8.8.1 Cable tray

Cable tray is available in a range of widths from 50 to 900 mm. The height of the sides is not defined by BS EN 61537[24] and manufacturers choose the height to provide the required strength. The sides may be a simple upstand, have a horizontal lip, or be turned through 180°, known as a return flange. The bottom of the cable tray may be solid or perforated.

Cable trays are manufactured from a variety of materials. Galvanised mild steel is by far the most common but glass reinforced plastic is frequently used where corrosion may be a problem.

Multicore cables can be laid on the base of the tray and, in most installations, need only be strapped in place with metal or plastic ties. Single core cables may need to be installed using cleats or clamps to prevent movement in the event of a short circuit.

Where cable trays are mounted in external locations it is recommended that a cover is installed to protect the cables from direct sunlight.

Many manufacturers use terms such as 'medium duty' and 'heavy duty' to describe their different products but these terms are not defined in the standard. Cable tray must be selected to carry the weight of the installed cables.

8.8.2 Cable basket

This was originally aimed at data cabling installations but has become popular for both small power and sub-main distribution systems. One of the main advantages over conventional tray is that cables can emerge in any direction. Also, cables can be routed through the side of the basket rather than over the flange of the tray, which can cause deformation or damage to the cable sheath.

Basket tray is typically available in widths from 50 to 600 mm with heights from 25 to 100 mm but these dimensions are not defined by the standard and will vary between manufacturers.

Most manufacturers can supply conduit mounting brackets so that conduit can be anchored firmly to the tray. This provides an easy route for connection to equipment or for drops to switches and socket-outlets. Dividers can be installed on the base of the tray to provide segregation between different types of circuit.

Basket tray is not really suited to the installation of single core power cables because there is no easy way to install the clamps onto the basket wires.

8.8.3 Cable ladders

Cable ladders are best suited to the installation of larger power cables. Widths are available from about 100 mm up to 600 or 900 mm for very heavy duty ladders. On most ladders the rungs are all the same height but some manufacturers produce ladders where alternate rungs are of different heights. This can cause a problem when calculating the loading because the weight is effectively carried by only half the rungs.

Designers and specifiers often require the installer to provide a proprietary cable cleat on every rung but it is debatable whether this is necessary for multicore cables, which can be secured with metal or nylon cable ties. However single core cables should be adequately restrained to prevent movement during short-circuit faults and this will probably require the cables to be installed in cleats.

8.8.4 Drop rods and hangars

This is applicable both to the cable support systems described above and to trunking systems.

The simplest method is to install threaded studding directly into the soffit and mount the tray or trunking on a steel channel. A basic installation has no lateral support and the system is able to move or swing from side to side. However this does not seem to be a problem in practice.

Where the system must be completely secured then restraining brackets or ties can be used.

One major disadvantage of this type of installation is that it is not possible to install the cables from the side, they must be threaded through the length of the installation. This does not occur where the tray or trunking is installed on brackets from a wall or hung on rigid brackets from the soffit.

8.8.5   Proprietary framing systems

These support systems are generally referred to as 'unistrut' systems, after the name of a manufacturer. Such systems are assembled on site from U-section channel and a relatively small number of connection components; they do not require any drilling or welding.

Cable trays and trunking can be supported from the soffit or cantilevered away from the wall. Access is available along the length of the run so cables can be installed from the side and do not need to be threaded from one end through all the drop-rods.

Where cables are installed vertically on walls it is usual for the channel to be fixed to the wall and the cables cleated directly to it.

This type of framing system is also used extensively for mounting switchgear and components.

8.9   Busbar trunking systems

8.9.1   General

Busbar trunking is basically a system of rigid conductors mounted on insulators inside a protective enclosure. Busbars are manufactured in a range of current ratings, typically 32 A or 63 A for use in final distribution systems, and up to several thousand amps for use in power distribution.

Conductors are generally copper in the smaller ratings but may be copper, aluminium or copper clad aluminium for the high current ratings. Some busbar conductors are sleeved with insulating material. The enclosure may be plastic or metal for the lower ratings but is normally metal for the higher ratings, often with ventilation openings. The current rating usually differs according to the busbar orientation, that is, whether the ventilation openings are at the top and bottom or on the sides.

Cast resin encapsulated busbars are available for use in corrosive atmospheres.

Busbar systems are generally intended for use on three-phase systems. Most manufacturers can offer a choice of neutral conductor sizes, ranging from half the phase conductor rating for use on balanced systems up to twice the phase conductor rating where heavy harmonic currents are anticipated. The CPC (earth) may be in the form of a dedicated fifth conductor or the metal enclosure may be used.

8.9.2   Power distribution

High current busbars are typically used for connections between transformers or generators to the main switchgear or for rising mains. In the latter case the busbar will have the facility to attach plug-in tap-offs for the connections to the floor distribution system. Tap-off points are provided at intervals of 300 mm to 600 mm.

Busbars are manufactured typically in sections of between 2 and 3 m in length. Connections between busbar sections vary greatly between manufacturers and in the simplest systems the busbar ends may be simply bolted together. More complicated connection methods include single or multiple bolts that clamp the ends of the busbars together. The clamping pressure is critical and manufacturers may require a torque wrench or supply double headed torque bolts that shear off when fully tightened.

The connections between the sections are the most common source of failure in busbar systems. Any dirt or foreign matter on the contact surfaces will result in a high impedance connection that will eventually overheat and fail. Installers need to take great care to ensure that the contact surfaces are completely uncontaminated and this may not be easy with other building work going on.

8.9.2.1   Busbar routes and connections to equipment

Manufacturers' catalogues frequently show examples of complicated busbar systems connected directly between items of equipment and even provide squared paper to assist in the design of such systems. Busbar sections are then specially manufactured in non-standard lengths to suit the proposed installation.

This approach may be suitable where the equipment is already installed and accurate dimensions are available but should be treated with caution for a new installation. The tolerances in normal building construction and the exact location of equipment will probably be much greater than the adjustment available in the assembled busbar system and it is quite likely that the connections will not mate up, particularly where rigid connections are used. In such cases the final connections should be made with flexible connections.

Flexible connections should also be used for connections to transformers. A busbar attached rigidly to a transformer will transmit the 100 or 120 Hz buzzing noise over a considerable distance.

8.9.2.2   Tap-off units

Most modern busbar systems employ plug-in tap-off units and with many designs these can be installed or removed without the need to de-energise the busbar. The way the tap-off unit connects to the busbars and the method of securing the tap-off unit vary widely but all systems will have some form of bolts or clamps to ensure that the unit cannot come loose.

Tap-off units may incorporate fuses, switch-fuses or circuit breakers according to the individual manufacturer's range. Some manufacturers produce both single and three-phase units, others only produce three-pole versions, particularly in the higher current ratings.

The connection from the tap-off unit to the distribution system should be either a multicore cable or single core cables in flexible conduit. All busbar systems will exhibit some movement due to expansion and contraction so rigid connections should not be used.

8.9.2.3 Feed units and accessories

End feed and centre feed units are available for installations where busbars are supplied via a cable. These normally bolt directly onto the busbars.

Busbar systems must allow for expansion due to temperature rise within the busbar itself and where the system crosses a building expansion or movement joint. All manufacturers cater for these with expansion couplings or straight sections incorporating suitable flexible components. On average busbar expansion joints are required at about 30 m intervals.

Larger busbar systems require a thrust block at the bottom of a vertical run and above any expansion joints. With some busbar types the thrust block is the only place where the busbar has any vertical support. The other mounting brackets are designed to allow the busbar to slide as it expands and contracts.

8.9.3 Underfloor systems

Underfloor busbars can provide a very flexible system for offices where the furniture layout is subject to frequent alteration. Busbars are installed under the raised floor and provide a supply to boxes installed in the floor tiles. Busbars from different manufacturers vary in rating but 32 A and 63 A ratings are the most common.

A typical 63 A system will incorporate a socket for connection of a tap-off unit at intervals of 300 mm. Most manufacturers can provide a variety of tap-off units but the two most commonly found are a fused unit incorporating a standard 13 A fuse and an unfused unit rated at 32 A. The unfused unit is normally used to connect to a floor box with two 13 A sockets.

The Wiring Regulations[1] limit the cable length of an unfused unit to a maximum of 3 m, so busbars are generally installed at about 6 m centres to give flexibility when locating the floor boxes.

Trunking systems are available in both single phase and three-phase versions. Single phase tap-off units for the three-phase systems usually provide a means of selecting the required phase. Trunking is also available with separate protective earth (CPC) and functional earth conductors and even with two electrically separate single phase systems.

Systems are available that comply with the Wiring Regulations section 607 requirements for high integrity protective connection requirements, used where equipment has high earth leakage currents.

The electrical supply to the trunking is connected via an end feed unit, generally wired in multicore cable. Trunking is manufactured in a variety of lengths up to about 3 m, but this varies between manufacturers. Individual lengths of trunking are connected together to form a rigid system although some manufacturers can supply flexible interconnections.

The term 'electrak' is often used as a generic name for underfloor trunking systems, after the name of a manufacturer.

8.9.4 Ceiling systems

Trunking systems for lighting control provide a very cost-effective solution in a modern office installation. The most basic systems are similar to those described for underfloor small power distribution where connections to the individual luminaires plug into switched conductors in the trunking. The manufacturers can provide prefabricated modular connections, so the high labour costs involved with a hard-wired system are avoided.

These types of systems can be very sophisticated and can incorporate all manner of control systems as an alternative to directly switched luminaires.

8.10 Prefabricated wiring systems

8.10.1 General

Prefabricated or modular wiring systems have become increasingly popular over the past few years. In principle, all the components can be manufactured in the factory and simply plugged together on site, thus resulting in a reduction in labour costs. There are additional advantages in that the system components can all be tested before delivery to site.

The types of components and cabling vary greatly between manufacturers and it is only possible to provide a general overview of the concept.

These types of system differ from the traditional conduit and trunking systems in that there is no containment and the pre-wired cabling is not enclosed in any way. The cables are often laid directly on the floor or on top of the ceiling grid. Several manufacturers use metal clad (MC) flexible cables to provide a degree of protection against damage.

8.10.2 Small power systems

The basic concept is similar to underfloor trunking, see above, except that the lengths of trunking are replaced by a network of connection boxes, sometimes referred to as zone boxes. These boxes have a fixed plug for incoming power and multiple output sockets. The boxes are connected together by a prefabricated cable with a plug on one end and a socket on the other. Supply cables to floor outlet boxes or socket-outlets on desks simply plug into the connection box.

Various current ratings are available, with 20 A and 32 A being common. One manufacturer's product uses 6 or 10 mm^2 conductors for the connection and supply cables and 4 mm^2 for tap-offs. Interconnection cables and tap-offs are available pre-wired in a range of lengths from around 3 to 10 m.

As with underfloor busbars, modular wiring systems are available with different combinations of protective and functional earth conductors.

8.10.3   Lighting systems

The principles are similar to those for underfloor systems. Some manufacturers can supply complete systems that incorporate lighting controls, supplies to HVAC plant such as fan coil units and connections back to a building management system (BMS).

8.11   Emergency and life safety systems

8.11.1   General

The term 'life safety' is not defined by the Wiring Regulations[1] but is frequently used to describe equipment such fire detection and alarm systems, firefighting pumps and lifts and the emergency lighting installation.

Many of these subjects are covered by the relevant CIBSE Guides and British Standards. This section covers only the way these relate to the building wiring.

8.11.2   Cable selection

8.11.2.1   Fire detection and alarm systems

Suitable cable types are listed in BS 5839-1[25]. In general the choice is between fire-resistant LSF/LS0H cables or MI cables, although the standard does include additional test requirements for certain cable types.

The standard requires that the methods of cable support do not reduce the circuit integrity below that afforded by the cable itself. They must be capable of withstanding a similar temperature as the cable and for the same length of time. This effectively rules out the use of plastic cable clips, cable ties or trunking.

The standard also recommends that cables be installed without joints, and that any unavoidable joints should also be capable of withstanding a similar temperature as the cable and for the same length of time. It would be difficult to make a joint satisfying these criteria, particularly with the LSF/LS0H types of cable.

There are also requirements for providing mechanical protection for the fire alarm cables but these are met by sensible installation practice such as additional protection to guard against specific risks, e.g. forklift trucks.

8.11.2.2   Emergency lighting systems

The current version of BS 5266-1[26] allows many different cable types and is similar in this respect to the previous version of the fire alarm standard BS 5839-1[25].

The standard is less prescriptive than the fire alarm standard although the general requirements for the integrity of the installation are similar. The standard also recommends that cables be routed through areas of low fire risk. It does not regard the wiring to self-contained emergency luminaires to be part of the emergency lighting circuits.

8.11.2.3   Firefighting lifts

Firefighting lifts in the UK are currently covered by British and European Standard BS EN 81-72[27] although the older BS 5588-5[28] is more specific in describing the electrical supply requirements.

According to BS 5588-5, a firefighting lift requires two separate supplies with an automatic changeover device located within the firefighting shaft, normally an automatic changeover switch in the lift machine room. The supplies need to be protected against the effects of the fire, either by installing them in the lift well, protecting them against the action of fire for a period not less than the structural fire protection of the firefighting shaft, or installing cables with a CWZ classification according to BS 6387[10]. The cable supports should be capable of surviving the effects of the fire.

Although not specifically required by BS 5588-5, the two supply cables should take different routes to the machine room.

8.11.2.4   Sprinkler system pumps

The current British Standard, BS EN 12845[29], requires that cables be protected against fire and mechanical damage and (a) run outside the building, or (b) run through parts of the building which are separated from any significant fire risk by walls, partitions or floors having fire resistance of not less than 60 minutes, or (c) are given additional protection, or (d) are buried.

This is less specific than the previous British Standard, BS 5306-2[30], which required that motor supply cables inside buildings had an AWX or SWX classification according to BS 6387[10], and were protected from direct exposure to fire or that they were mineral insulated.

8.12   Cable routes within the building

8.12.1   Mains and sub-mains

Most modern multi-storey buildings will have one or more service cores with dedicated electrical riser cupboards on each floor. The main incomers, whether high or low voltage, will probably be at ground floor or basement level. The main switchgear should be located as close as possible to the risers to avoid excessively long cables running horizontally in the basement to the bottom of the risers. If the risers are very far apart it may be better to have separate incomers and switchgear for each riser if this can be arranged.

8.12.1.1 High level installation

The best routes for the sub-main cables is along a circulation route or corridor. In most basements the corridors will be high enough to allow the cables to be installed at high level on cable trays or ladders either suspended from the soffit or mounted on the walls. Cable routes between the main switchrooms and the risers should be as direct as possible but cables should not be routed through mechanical plantrooms unless this is unavoidable and should never be routed through boiler rooms as these can obviously become very warm. It is also advisable to avoid areas where cables may be more prone to damage, such as car parks and loading bays.

In many cases the support system for the cables can be shared with the mechanical services but this may only be practicable when both the mechanical and electrical installations are carried out by the same contractor. Care must be taken where cables are installed in close proximity to hot water or steam services to ensure that overheating cannot occur, although in most cases the sub-mains will have been sized on voltage drop and a slight reduction in current carrying capacity may be acceptable.

8.12.1.2 Trench systems

The alternative to installation at high level is to install the cables in trenches cast into the floor construction. Cables may be laid directly on the floor of the trench but this may result in damage if they are submerged in water for extended periods of time. A more satisfactory arrangement is to install proprietary brackets on the side of the trench or to use cable ladder or tray.

Cables installed in a covered trench will need to be derated because there is very little airflow to dissipate the heat. Derating factors are given in the Wiring Regulations[1] and cables may need to be derated by as much as 50% under extreme circumstances. Cable tunnels on large industrial projects are normally designed with ventilation systems to dissipate the heat but this is impractical for building installations.

Where trenches are interconnected by ducts it is advisable to install spare ducts to allow for future additions and alterations.

Trenches should be fitted with covers that seal against the ingress or water and dirt as much as possible, although there will never be a perfect seal. Trench covers must be strong enough to carry the weight of any likely loads, particularly heavy plant that may need to be moved along basement passageways.

8.12.2 Reliability and redundancy

Modern financial and data processing installations require a very secure power supply and these normally incorporate duplicated supply cables and redundant or standby switchgear.

It is imperative that the duplicated supply systems are separated physically as well as electrically. In many installations both the cables are installed in the same electrical riser. An incident such as a fire in the riser cupboard would result in both cables being damaged and the critical supply being lost. The duplicated cables should follow completely separate routes right up to the final changeover or transfer switchgear.

8.13 Installations in particular building types

The installation methods discussed so far apply to most types of domestic, commercial and industrial buildings. In most countries the requirements for such installations are covered in the national wiring regulations and codes, such as the Wiring Regulations[1].

However other types of building, particularly those accessible to the general public, may be subject to additional approvals or regulations. Many of these regulations are determined at local rather than a national level and may be different in neighbouring towns, so it is not possible to give details of the exact requirements.

In England and Wales all theatres and other places licensed for public entertainment, music, dancing etc. are subject to the Local Government (Miscellaneous provisions) Act[31]. In Scotland, the Civic Government (Scotland) Act[32] applies. Local authorities dictate the exact requirements and typically these may include provision of residual current protection on socket-outlets and more frequent periodic inspection and testing. Cinemas are covered by various acts such as the Cinematograph (Safety) Regulations[33].

Where these or similar situations are encountered it is essential to liaise with the relevant local authority such as the Building Control Officer or District Surveyor.

8.14 Hazardous areas

This is a specialised field and only the main principles can be given here. Hazardous areas are classified by BS EN 60079: Part 10[34] as one of three zones, based on the probability that hazardous conditions are present. Explosive gases are classified into three groups according to their flammability.

All equipment installed in a hazardous area must be certified by an appropriate body and marked accordingly. The markings indicate where the equipment may be installed, according to the zone and gas classification. Equipment such as luminaires that will become warm in normal use have an indication of the maximum surface temperature. In Europe the certifications is governed by the ATEX Directive[35].

There are various methods for ensuring the suitability of equipment for use in a hazardous area, referred to as the protection type, and defined in the various parts of BS EN 60079[34]. These definitions, such as 'increased safety', are mainly of concern to the manufacturer. The designer or specifier is more concerned that the equipment is certified for the hazard, rather than how the certification was achieved.

Installation in a hazardous area can be carried out in single core cables in conduit or with multicore cables. In the UK and Europe the use of multicore cables has now become more popular because of the generally simpler installation methods. Basically the installation is the same as a conventional installation except that all equipment and cable glands are certified products.

8.15   Mains signalling

Several systems have been produced for sending control or data signals over the mains wiring. These generally have specific requirements to enable them to work correctly, e.g. transmitters and receivers may need to be connected to the same phase of a three-phase system. Consequently it may not be possible to install these into an existing installation. Where the use of such systems is envisaged it will be necessary to discuss the precise details with the manufacturer at the early design stages.

8.16   Electromagnetic compatibility

8.16.1   Electromagnetic compatibility at 50/60 Hz

All electrical currents generate a magnetic field. In most parts of an installation the phase and neutral conductors are in close proximity and the individual magnetic fields will tend to cancel each other out. However in some areas the cables may not be close to each other, or may be installed in an asymmetric pattern. This is often the case where a supply transformer is connected to the main switchboard with single core cables and the high currents can result in strong magnetic fields. These magnetic fields may be strong enough to affect cathode ray tubes in computer monitors or televisions and render the areas close to the switchroom unusable for normal office work. These stray fields can usually be avoided by ensuring that the cable installation is as close to symmetrical as possible, although in some cases it may be necessary to screen the cables by enclosing them in steel trunking, see section 8.7.1.

A typical computer monitor will be affected by a field strength of about $1\ \mu T$ and unusable at about $10\ \mu T$. For comparison, the National Radiological Protection Board (NRPB) recommends a limit of 1.6 mT as the maximum continuous exposure for humans.

In most mains frequency (50/60 Hz) systems, the magnetic fields will not cause interference to data or communication systems provided that such circuits are separated by a minimum of 300 mm.

8.16.2   Electromagnetic compatibility and harmonics

A more serious problem occurs when there are harmonics in the system. The cable can act as an antenna and radiate these higher frequencies, which are then picked up by other cables.

One of the main sources has been the variable speed drives commonly used on mechanical plant. Manufacturers can provide advice on how to avoid excessive radiation, sometimes recommending screened rather than armoured cables for the final connection to the motor. The connection and routing of the motor earth cable is sometimes critical.

Electromagnetic compatibility is considered in detail in section 11.

8.17   Earthing

See also section 10 for detailed consideration of earthing.

All circuits in a fixed installation require a protective earth conductor, correctly known as the circuit protective conductor or CPC. All exposed-conductive-parts of the installation (the hyphens are part of the IEE definition) that may become live under fault conditions must be connected to the CPC. Typical examples of exposed-conductive-parts include metal switchboard enclosures, steel containment and support systems, luminaire bodies, motor frames etc. In the event of such parts becoming live the current will return to the source via the CPC and be of sufficient magnitude to operate the protective device.

All exposed conductive parts are connected to the CPC, and consequently to each other, so under an earth fault condition the voltages between simultaneously accessible parts will not be of such magnitude and duration to be dangerous. This type of protection is known as 'earthed equipotential bonding and automatic disconnection of supply' (EEBADS).

In theory, practically any metallic path can be used as the CPC but some are better than others as detailed below.

8.17.1   Earth loop impedance

In the event of contact between a live conductor and an exposed conductive part, the current will flow down the live conductor and back to the source through the CPC. The magnitude of this current must be sufficiently large to operate the upstream protective device within a short time. In the UK the Wiring Regulations[1] require that the protective device operates within five seconds where the circuit supplies fixed equipment or 0.4 seconds where it supplies socket-outlets.

The exposed-conductive-part will become live, at a potential determined by the relative resistances of the supply conductor and the return CPC conductor. In the case of a portable appliance connected via a plug and socket, these two resistances will probably be of similar value and the appliance will reach a potential of around half the live conductor voltage.

The current that will flow is determined by the voltage and the overall impedance of the circuit along the live conductor and back up the CPC, including any impedance of the external connections and the source itself. This impedance is known as the earth fault loop impedance.

The Wiring Regulations include tables of maximum earth loop impedance for commonly used types and ratings of protective devices.

8.17.2 Steel trunking and conduits

Steel conduit and trunking can be used as the sole CPC, see section 8.7.1) but extra care is required on the part of the installer to ensure that there is a sound electrical connection between the component parts. This is simple where galvanised conduits are screwed together but poses problems when connecting conduit to painted trunking. The paint must be removed from the trunking surface to get good metal to metal contact. However, this removes the protection and makes the metal more likely to rust, which may result in the loss of electrical continuity.

A problem also occurs where additions or alterations are made to the installation. It is not possible to cut or remove sections of the trunking while any of the circuits are live because this would disconnect the CPC from the downstream system.

For these reasons, many designers and specifiers do not permit conduit and trunking systems to be used as the CPC and insist that dedicated conductors are installed.

8.17.3 Installations using single core cables

Where plastic trunking and conduit systems are used, or where steel containment systems are not allowed to be used as the CPC, it is usual to install a single core CPC cable with the circuit phase and neutral conductors. On final sub-circuits this CPC cable is often the same cross sectional area as the phase and neutral, although the calculations will usually permit one size smaller.

In some cases it is possible to use a single CPC cable for more than one circuit. For example, a multi-gang switch box controlling several lighting circuits would only require a single CPC, not one for each circuit. This is not the case with small power ring circuits as the CPC must be run in a ring with the phase and neutral conductors.

8.17.4 Multicore armoured cables

Four-core armoured cables are used extensively in the UK on three-phase and neutral systems, with the armouring used for the CPC. With long cable runs it is often found that the earth fault loop impedance is too high to ensure that the protective device will operate within the five second period. The solution is simply to install an additional single core cable in parallel.

However a degree of care needs to be taken when determining the size of this additional CPC. Some electrical design software will indicate that only a very small CPC is required to reduce the earth fault loop impedance to an acceptable value. Sometimes cables with cross sectional areas as small as 4 or 6 mm^2 are indicated, particularly where the disconnection time without the additional CPC is only just over five seconds.

Installing cables of this size in parallel with large power cables looks very strange, particularly when the bonding conductors on the support systems are much larger, and may generate queries from the installation contractor.

8.17.5 Support systems

In the UK it is permissible to use an electrically continuous support system as the CPC. Most galvanised steel cable tray and ladder systems will be firmly bolted together and there will be good continuity across the joints. Normally bonding conductors will be installed between separate sections or across building expansion joints.

However this type of CPC is very unusual and it would be difficult to ensure that no alterations or modifications reduced the efficacy of the earth path. It is not recommended that this solution be used in building installations.

8.17.6 Equipotential bonding

The main danger to persons is if they are in simultaneous contact with conductive parts that are at different potentials.

The Wiring Regulations[1] define 'extraneous-conductive-parts' which are conductive parts not forming part of the electrical installation. These include non-electrical services such as pipes and ducting. Such services may introduce earth potential alongside the electrical installation and the equipment served by it, making it quite likely that a person could be in simultaneous contact with an exposed-conductive-part made live by an earth fault and a nearby extraneous-conductive-part.

To avoid this situation it is normal to install bonding conductors between these different extraneous-conductive-parts and between the extraneous-conductive-parts and the exposed-conductive-parts.

This equipotential bonding is to be found in all sizes and types of installation from small domestic up to large commercial and industrial installation.

Equipotential bonding is considered in detail in section 10.3.10.

8.18 Testing and inspection

In the UK the Wiring Regulations[1] give full details of the inspection and testing requirements for new installations and for the periodic inspection of existing installations.

Testing and inspection is considered in detail in section 12 of this Guide.

8.18.1   Visual inspection

The Wiring Regulations[1] require good workmanship for all installations, but this is a very subjective quantity. A visual inspection will provide a good indication as to the general standard of workmanship and may indicate where the installer has taken shortcuts or produced an unacceptable finished product, even if these do not specifically contravene any of the Regulations.

8.18.2   Certification

All new installations in the UK must be subjected to the inspection and test procedures detailed in the Wiring Regulations[1]. After completion of this work a certificate must be issued to the person who ordered the installation work and this certificate must include full test results as well as any deviations or departures.

Examples of suitable certificates are included in the Wiring Regulations. The full certificate includes separate sections where the relevant parties can certify the design, installation and testing separately, as these will invariably be done by different persons on a large installation. There is also a more simple certificate to be used where a single person has responsibility for all three areas, as may be the case with a domestic installation.

8.18.3   Periodic inspection and testing

The Wiring Regulations[1] require installations to be subject to a periodic inspection and testing regime, but the requirements have changed over the years. Previous editions stated that the installation should be tested at a maximum interval of five years, or a shorter interval if recommended by the person doing the testing.

The current requirements are less specific and allow the testing to be integrated with a planned preventative maintenance routine, provided that suitable records are kept. In some cases this would allow a part of the installation to be tested and the results compared with those obtained previously. If no deterioration had occurred then it may be deemed unnecessary to test the remainder. In reality this is probably an acknowledgement of the much greater life expectancy of modern cables and insulation materials.

8.19   Guidance for designers and specifiers

Building wiring systems in commercial installations have changed considerably over the past few years although domestic and industrial installations have remained much the same.

8.19.1   Domestic installations

In the UK the majority of installations are still carried out using flat twin and earth type wiring, although this may now be LSF or LS0H rather than PVC/PVC.

In two-storey dwellings with a timber first floor it is commonplace to run all the small power ring main wiring under the first floor, dropping down to the socket-outlets on the ground floor. Lighting circuits for the ground floor are also run under the floor while lighting circuits for the upstairs rooms are run in the attic or loft space.

The vertical cable runs to socket-outlets and switches are normally given additional protection by a PVC channel but may be buried in the plaster. The channel does not normally allow the cable to be replaced, or more cables to be added.

In exposed parts of the installation, such as in a garage, cables may be directly mounted onto the surface or may be installed in plastic conduit or mini-trunking. The latter provides some additional protection and gives the impression of a more professional job.

8.19.2   Industrial installations

The type of installation should be selected according to the degree of mobility of equipment or flexibility required.

Final circuits supplying machine tools or equipment subject to frequent relocation are best served by using plug-in busbar trunking run at high level. Where equipment is static it is common in the UK to use single core cables in steel conduit and trunking. The same principals can be applied to the lighting installation. European practice will commonly use multicore non-armoured cables.

Sub-mains to local distribution boards can be run in armoured multicore cables installed at high level on cable trays or ladders.

8.19.3   Office and commercial installations

The modern office installation seems to have a typical lifespan of only a few years before being extensively redesigned and flexibility and ease of alteration are now the key points.

Traditionally these installations have been carried out using single core cables in steel conduit and trunking for both the lighting and small power installation. This type of installation is very difficult to modify to suit a change in layout and in many cases the original containment systems may simply be left in place when a refurbishment is carried out.

Where an office has a raised floor the small power circuits can be supplied using 32 A or 63 A plug-in busbar trunking. This provides the flexibility to locate or move the socket-outlets to convenient locations for desks and office furniture. Where there is no raised floor it is possible to install proprietary poles incorporating the socket-outlets and any communication and data services. These can be supplied from similar busbar trunking run above the suspended ceiling. This type of installation is often deemed to be unacceptable on aesthetic grounds but may be the only solution for large open office spaces where there is no possibility of access from below.

Lighting circuits can also be supplied via plug-in busbar trunking systems that can incorporate all the switching and control functions. Final connections to the luminaires may simply be flexible cords run above the suspended ceiling, provided that there is no danger of mechanical damage to the cables. An alternative is to use a completely modular system with plug-in cables and connectors. This has the advantage that the individual items can probably be re-used to suit further changes in luminaire layout.

Whichever solution is adopted it should provide the required degree of flexibility and this effectively rules out the steel containment systems that have served for so long.

References

1   BS 7671: 2001: *Requirements for electrical installations. IEE Wiring Regulations. Sixteenth edition* (London: British Standards Institution) (2001)

2   *Selection and Erection of Equipment* Guidance Note 1 to IEE Wiring Regulations (London: Institution of Electrical Engineers) (1999) ISBN: 0852969546

3   BS 1363: *13 A plugs, socket-outlets and adaptors*; Part 1: 1995: *Specification for rewirable and non-rewirable 13 A fused plugs*: Part 2: 1995: *Specification for 13 A switched and unswitched socket-outlets*; Part 3: 1995: *Specification for adaptors*; Part 4: 1995: *Specification for 13 A fused connection units switched and unswitched* (London: British Standards Institution) (1995)

4   *Isolation and Switching* Guidance Note 2 to IEE Wiring Regulations (London: Institution of Electrical Engineers) (1999) ISBN: 0852969554 1999

5   *Protection Against Fire* Guidance Note 4 to IEE Wiring Regulations (London: Institution of Electrical Engineers) (1999) ISBN: 0852969570

6   *Protection Against Electric Shock* Guidance Note 5 to IEE Wiring Regulations (London: Institution of Electrical Engineers) (1999) ISBN: 0852969589 1999

7   *Protection Against Overcurrent* Guidance Note 6 to IEE Wiring Regulations (London: Institution of Electrical Engineers) (1999) ISBN: 0852969597

8   BS 6360: 1991: *Specification for conductors in insulated cables and cords* (London: British Standards Institution) (1991)

9   BS EN 60702: *Mineral insulated cables and their terminations with a rated voltage not exceeding 750 V*: Part 1:2002: Cables (London: British Standards Institution) (2002)

10 BS 6387: 1994: *Specification for performance requirements for cables required to maintain circuit integrity under fire conditions* (London: British Standards Institution) (1994)

11 BS 7629: *Specification for 300/500 V fire resistant electric cables having low emission of smoke and corrosive gases when affected by fire*: Part 1: 1997: *Multicore cables*; Part 2: 1997: *Multipair cables* (London: British Standards Institution) (1997)

12 BS 7846: 2000: *Electric cables. 600/1000 V armoured fire-resistant cables having thermosetting insulation and low emission of smoke and corrosive gases when affected by fire* (London: British Standards Institution) (2000)

13 BS EN 50200: 2000: *Method of test for resistance to fire of unprotected small cables for use in emergency circuits* (London: British Standards Institution) (2000)

14 BS 8434: *Methods of test for assessment of the fire integrity of electric cables. Test for unprotected small cables for use in emergency circuits*: Part 1: 2003: *BS EN 50200 with addition of water spray*; Part 2: 2003: *BS EN 50200 with a 930 °C flame and with water spray* (London: British Standards Institution) (2003)

15 BS EN 60702: *Mineral insulated cables and their terminations with a rated voltage not exceeding 750 V*: Part 1: 2002: *Cables*; Part 2: 2002: *Terminations* (London: British Standards Institution) (2002)

16 BS 6004: 2000: *Electric cables. PVC insulated, non-armoured cables for voltages up to and including 450/750 V, for electric power, lighting and internal wiring* (London: British Standards Institution) (2000)

17 BS 7211: 1998: *Electric cables. Thermosetting insulated, non-armoured cables for voltages up to and including 450/750 V, for electric power, lighting and internal wiring, and having low emission of smoke and corrosive gases when affected by fire* (London: British Standards Institution) (1998)

18 BS 7889: 1997: *Electric cables. Thermosetting insulated, unarmoured cables for a voltage of 600/1000 V* (London: British Standards Institution) (1997)

19 BS 5467: 1997: *Electric cables. Thermosetting insulated, armoured cables for voltages of 600/1000 V and 1900/3300 V* (London: British Standards Institution) (1997)

20 BS 6724: 1997: *Electric cables. Thermosetting insulated, armoured cables for voltages of 600/1000 V and 1900/3300 V, having low emission of smoke and corrosive gases when affected by fire* (London: British Standards Institution) (1997)

21 BS 6500: 2000: *Electric cables. Flexible cords rated up to 300/500 V, for use with appliances and equipment intended for domestic, office and similar environments* (London: British Standards Institution) (2000)

22 BS 7919: 2001: *Electric cables. Flexible cables rated up to 450/750V, for use with appliances and equipment intended for industrial and similar environments* (London: British Standards Institution) (2001)

23 BS EN 60947:*Specification for low-voltage switchgear and controlgear*: Part 1: 1999: *General rules* (London: British Standards Institution) (2001)

24 BS 4678: *Cable trunking*: Part 1: 1971: *Steel surface trunking*; Part 4: 1982: *Specification for cable trunking made of insulating material* (London: British Standards Institution) (1971, 1982)

25 BS EN 61386: *Conduit systems for cable management*: Part 1: 2004: *General requirements*; Part 21: 2004: *Particular requirements. Rigid conduit systems*; Part 22: 2004: *Particular requirements. Pliable conduit systems*; Part 23: 2004: *Particular requirements. Flexible conduit systems* (London: British Standards Institution) (2004)

26 BS 4568: *Specification for steel conduit and fittings with metric threads of ISO form for electrical installations*: Part 1: 1970: *Steel conduit, bends and couplers* (London: British Standards Institution) (1970)

27 BS EN 61537: 2002: *Cable tray systems and cable ladder systems for cable management* (London: British Standards Institution) (2002)

28 BS 5839: *Fire detection and alarm systems for buildings*: Part 1: 2002: *Code of practice for system design, installation, commissioning and maintenance* (London: British Standards Institution) (2002)

29 BS 5266: Emergency lighting: Part 1: 1999: *Code of practice for the emergency lighting of premises other than cinemas and certain other specified premises used for entertainment* (London: British Standards Institution) (1999)

30 BS EN 81: *Safety rules for the construction and installation of lifts. Particular applications for passenger and goods passenger lifts*: Part 72: 2003: *Firefighters lifts* (London: British Standards Institution) (1999)

31 BS 5588: *Fire precautions in the design, construction and use of buildings*: Part 5: 1991: *Code of practice for firefighting stairs and lifts* (London: British Standards Institution) (1991)

32   BS EN 12845: 2003: *Fixed firefighting systems. Automatic sprinkler systems. Design, installation and maintenance* (London: British Standards Institution) (2003)

33   BS 5306: *Fire extinguishing installations and equipment on premises*: Part 2: 1990: *Specification for sprinkler systems* (London: British Standards Institution) (1990) (superseded)

34   Local Government (Miscellaneous Provisions) Act 1976 (London: Her Majesty's Stationery Office) (1976)

35   Civic Government (Scotland) Act 1982 Chapter 45 (London: Her Majesty's Stationery Office) (1982)

36   The Cinematograph (Safety) Regulations 1955 Statutory Instruments 1955 No. 1129 (London: Her Majesty's Stationery Office) (1955)

37   BS EN 60079: *Electrical apparatus for explosive gas atmospheres*; Part 10: 2003: *Classification of hazardous areas* (London: British Standards Institution) (2003)

38   Directive 94/9/EC of the European Parliament and the Council of 23 March 1994 on the approximation of the laws of the Member States concerning equipment and protective systems intended for use in potentially explosive atmospheres *Official J. of the European Communities* **L060** 14–15 (3/03/1994)

Bibliography

British Standards

The notes given below are intended to indicate the content of the standard where this is not completely clear from the title. Several of the standards listed consist of more than one part.

Cables

BS 6346: 1997: *Electric cables. PVC insulated, armoured cables for voltages of 600/1000 V and 1900/3300 V* (*Note*: covers pvc insulated, pvc sheathed power cables; editions up to 1997 also covered non-armoured cables.)

BS 5467: 1997: *Electric cables. Thermosetting insulated, armoured cables for voltages of 600/1000 V and 1900/3300 V* (*Note*: covers XLPE or EPR insulated, PVC sheathed power cables; editions up to 1997 also covered non-armoured cables.)

BS 6724: 1997: *Electric cables. Thermosetting insulated, armoured cables for voltages of 600/1000 V and 1900/3300 V, having low emission of smoke and corrosive gases when affected by fire* (*Note*: covers power cables with LSF/LS0H insulation and sheath, but not fire resistant.)

BS 7846: 2000: *Electric cables. 600/1000 V armoured fire-resistant cables having thermosetting insulation and low emission of smoke and corrosive gases when affected by fire* (*Note*: covers fire resistant power cables with LSF/LS0H insulation and sheath.)

BS 6004: 2000: *Electric cables. PVC insulated, non-armoured cables for voltages up to and including 450/750 V, for electric power, lighting and internal wiring* (*Note*: covers PVC insulated cables, unsheathed or PVC sheathed. Includes conduit wires and flat twin and earth.)

BS 7211: 1998: *Electric cables. Thermosetting insulated, non-armoured cables for voltages up to and including 450/750 V, for electric power, lighting and internal wiring, and having low emission of smoke and corrosive gases when affected by fire* (*Note*: covers similar cables to BS 6004 but with LSF/LS0H insulation and sheath.)

BS 7629: *Specification for 300/500 V fire resistant electric cables having low emission of smoke and corrosive gases when affected by fire*: Part 1: 1997: *Multicore cables*; Part 2: 1997: *Multipair cables* (*Note*: covers fire resistant cables with LSF/LS0H insulation and sheath.)

BS 6500: 2000: *Electric cables. Flexible cords rated up to 300/500 V, for use with appliances and equipment intended for domestic, office and similar environments*

BS 7919: 2001: *Electric cables. Flexible cables rated up to 450/750V, for use with appliances and equipment intended for industrial and similar environments*

BS EN 60702: *Mineral insulated cables and their terminations with a rated voltage not exceeding 750 V*: Part 1: 2002: *Cables*; Part 2: 2002: *Terminations*

Joints

BS EN 61238: *Compression and mechanical connectors for power cables for rated voltages up to 36 kV (U_m = 42 kV)*: Part 1: 2003: *Test methods and requirements* (*Note*: covers compression and mechanical connectors for cables larger than 10 mm^2.)

Trunking, conduits, trays, ladders

BS 4568: *Steel conduits*

BS 4678: *Cable trunking*: Part 1: 1971: *Steel surface trunking*; Part 2: 1973: *Steel underfloor (duct) trunking*; Part 4: 1982: *Cable trunking. Specification for cable trunking made of insulating material*

BS EN 50085: *Cable trunking and cable ducting systems for electrical installations*: Part 1: 1999: *General requirements*; Part 2: 2001: Section 3: *Particular requirements for slotted cable trunking systems intended for installation in cabinets* (*Note*: covers metallic and non-metallic trunking.)

BS EN 50086: *Specification for conduit systems for cable management*: Part 1: 1994: *General requirements*; Part 2: *Particular requirements*: Section 1: 1996: *Rigid conduit systems*; Section 2: 1996: *Pliable conduit systems*; Section 3: 1996: *Flexible conduit systems*; Section 4: 1994: *Conduit systems buried underground* (*Note*: covers metallic and non-metallic conduits; rigid, flexible and pliable.)

BS EN 61537: 2002 (IEC 61537: 2001): *Cable tray systems and cable ladder systems for cable management* (*Note*: covers cable tray and ladder systems, including basket tray.)

BS EN 60439: *Specification for low-voltage switchgear and controlgear assemblies*: Part 2: 2000 (IEC 60439-2: 2000): *Particular requirements for busbar trunking systems (busways)* (*Note*: covers busbar trunking systems.)

9   Uninterruptible power supplies

9.1   Introduction

The need for clean reliable power is increasing and is being met today by the provision of uninterruptible power supply (UPS) systems of varying technologies and capacities. Historically, UPS was provided only for such applications as airport lighting, communications systems or mainframe computers. Initially they were provided for short term duty, using flywheels to store energy to cover the time span between mains failure and restoration of supply from standby generators. Today the needs can be communications equipment and system processors up to large data centres with mixed processor and communications equipment etc., operating continuously with little, if any, opportunity for maintenance shut-downs. Medical monitoring systems and industrial process plants are other typical loads. The loads served have also changed dramatically with equipment using phase controlled rectifiers and switched mode power supplies, which themselves are more susceptible to power disturbances whilst reflecting load harmonic current which can cause supply problems.

The function of any UPS is two-fold:

—   to stabilise unstable supplies

—   to protect against complete loss of supply.

In terms of the stabilising function, the necessary output-to-load tolerances of (typically) ±1% in voltage and ±1% in frequency should be maintained, without the use of the battery (i.e. under normal UPS running conditions, with input supply variations of ±10% in voltage and ±5% in frequency).

Today's sensitive loads, whether data processing or communication systems, can generally accept a break or disruption in power supply for a time not exceeding 10 ms. In the range of 10 ms to a few seconds there are many deviations on the power supply (outside of the input tolerances for such equipment), which include:

—   frequency variations

—   voltage dips and spikes

—   micro-breaks in one or more phase

—   'brown-outs'

—   power surges

—   black-outs.

Therefore a UPS has to perform a vital role in power conditioning, protecting the load from the vagaries of the raw mains and other consumers from the harmonics generated by the critical load. Mains quality has an impact upon the overall system reliability and it is therefore important to have basic data upon which to make comparisons. Table 9.1 represents a long term study of 34

Table 9.1 Mean time between failures and mean down time for 34 commercial consumers during 1983

Measure	Value		
	Worst	Average	Best
Mean time between failures (h)	43	155	685
Mean down time (s)	8.145	1.72	0.1

commercial consumers during 1983 showing the mean time between failure (MTBF) and the mean down time (MDT) of the mains power supply for all deviations of voltage and frequency longer than 10 ms in duration. (See section 9.3.2 for consideration of system MTBF.)

Note that the MDT does not only indicate loss of power but also deviations to the specification for 'computer grade' power (typically ±1% in voltage and ±1% in frequency).

Where a UPS supplier or manufacturer quotes an MTBF figure for a module or system it must always be linked to the quality of the mains supply.

Having recognised the need for UPS there are a number of systems available from small, single phase modules produced to meet the needs of dedicated micro-computers, up to large multi-module 'redundant' systems serving distribution networks throughout large multi-function developments.

There are currently two basic designs of UPS, static transistor and rotary, although historically thyristor technology was used for larger, static systems. These systems are described later but it should be noted that the technology used influences both the location of equipment within a development and the way it is applied to the loads. System characteristics can themselves be considered in three categories:

—   dedicated 'in room' (i.e. within the office environment) units, up to 5 kV·A single phase

—   three-phase units up to 160 kV·A (which are designed for 'in room' installation but can only accommodate integral batteries up to about 30 kV·A)

—   larger systems ranging between 160 kV·A to many MV·A, which would require plant room accommodation.

9.2   Basic types of UPS systems

Historically, regardless of whether or not the inverter power switching devices were thyristors or transistors, the output stage is a transformer ('static UPS') or rotating generator ('rotary UPS'). Recent power switching technology has allowed the omission of these transformers.

The arrangement of the basic components falls into four configurations:

— off-line

— triport

— parallel on-line

— series on-line, or double power conversion.

Each configuration fundamentally changes the quality of power supply to the load, and the effect that the UPS/load has on the mains supply. It is important that the specifier can recognise the equipment type on offer by comparing operating performance.

In off-line systems (see Figure 9.1) the load is normally supplied from the mains via a static transfer switch. The rectifier acts as a battery charger when the mains is present. Mains failure causes the transfer switch to disconnect the load from the mains and connect it to the inverter. There can be a break in the supply in the order of 3 to 4 ms.

A triport system (see Figure 9.2) incorporates a combined inverter/charger capable of both recharging the battery when the mains is available and supplying the load with AC power when the static switch has disconnected the mains.

Both the off-line and triport designs provide limited power conditioning but high efficiency. They both take advantage of the fact that the power components need not be fully rated and are usually smaller and less costly.

Additional filtering adds to the cost and reduces both the efficiency and reliability of such systems. Due to the triport design's reliance upon the mains supply, it can be susceptible to problems when the incoming voltage and frequency tolerance band is wide, particularly when being supplied by a standby diesel generator. The effect is for the UPS to switch to battery operation and stay in that mode

until the battery cut-off voltage is reached and the UPS switches off.

Parallel on-line systems (see Figure 9.3) normally supply the load with raw mains via a static switch (for rapid disconnection of the mains) and, often, a series reactor. The power switching stage runs (off-load) in parallel with the mains, ready to take over the load in the event of mains failure. The normal running efficiency is therefore high but the degree of isolation of the load harmonics from the mains supply is limited, as is the protection against mains disturbances reaching the load. Savings in cost can be achieved due to the fact that the power switching elements in the inverter (rotating machine in the case of a rotary UPS) need not be fully rated.

Series on-line systems (see Figure 9.4) have a fully rated power path from the mains to the load, and process 100% of the power in normal operation. Modern designs, both static and rotary, offer full galvanic isolation (mains from load) with high efficiencies.

All of the above systems can be fitted with a maintenance bypass (although usually unnecessary for small off-line systems), and no-break auto-bypass facilities. Maintenance can therefore be carried out without a break in the supply to the load.

9.2.1 Dedicated range up to 5 kV·A

These units are generally for single phase operation and are manufactured in cabinets sufficiently small that they can be installed beneath desk tops. They are generally employed to support either single micro-computers or a file server for a small group of terminals in a localised area. Above this capacity, UPS systems are applied to provide more general support via power distribution units (PDUs).

There are many manufacturers producing units with various capacities in the range 100 V·A to 5 kV·A. This group of UPS modules can be further divided into 'on-line', 'off-line' and 'triport' categories.

On-line UPS systems are designed to condition the supply to the critical load continuously by the use of the normal power route, i.e. the inverter is working even when the

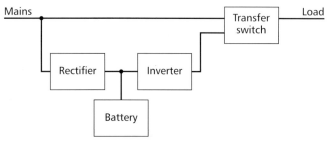
Figure 9.1 Off-line UPS system

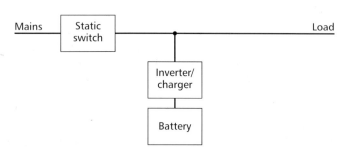
Figure 9.2 Triport UPS system

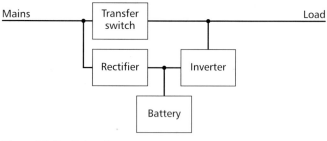
Figure 9.3 Parallel on-line UPS system

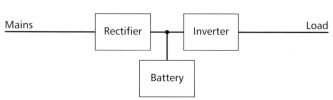
Figure 9.4 Series on-line UPS system

mains is healthy. The system uses a fully rated rectifier to convert the incoming AC mains into DC; this DC supply is then inverted via high speed switching devices into a synthesised AC supply. The batteries, housed within the UPS enclosure, are connected to and charged from the DC stage of the equipment. The majority of systems currently available use high speed transistor technology in the inverter to produce a pulse width modulated (PWM) output which can react to maximum current demands and maintain an approximately sinusoidal output voltage waveform.

Historically, thyristor technology was employed in the inverter stage to produce a stepped waveform which had a high harmonic content and had to be smoothed, using filters, to give an acceptable supply waveform. Due to the loads to which these units are being applied, and the greater availability of power transistors, the majority of units in this range now use transistor technology.

The units are interposed between the raw mains and the critical load. A 'healthy' mains condition is when the input voltage and frequency are within acceptable input tolerances, typically ±10% and ±5% respectively. Where the input supply wanders outside the given tolerances, the output supply is derived from the battery. If input supplies are frequently variable, or long term generator support is expected due to an unreliable supply, then consideration should be given to an increased battery autonomy period to maintain capacity in readiness for total supply failure.

Total output harmonic voltage distortion (THVD) levels of 5% are typical for these units (when supplying a linear load) but when non-linear loads with high crest factors are expected de-rating may be necessary if the THVD is to be kept low.

Component failure or short term overloads are always likely in any system and to protect the critical load from disconnection, automatic bypasses are employed in many systems. This bypass leg is served from the raw mains and is connected to the critical load output terminals by means of a static switch. Under component failure or overload conditions, and provided that the 'raw' mains supply is within tolerances of (typically) ±10% and ±1% for voltage and frequency respectively, the load will be connected to the bypass. It is usual that, once overload conditions have passed, the critical load will revert to the normal power route.

The overload conditions can include a downstream short circuit and this type of UPS, along with all such static systems incorporating an inverter and output transformer, has a relatively high output impedance (high sub-transient reactance). Therefore they have to resort to the mains supply to clear a fault. Both the actual transfer operation and availability of the mains introduces a risk element in the supply integrity.

In their basic form, systems generally comprise integral batteries, chargers, inverters, and general circuit and component protection with basic status indication provided on the enclosure. The units are constructed to operate between temperature limits of 0 °C and 40 °C, and are provided with acoustic attenuation such that noise emission under normal operation is below 60 dB(A) at 1 metre. This equipment can be augmented by 'plug-in' diagnostic indicators offering status and normal metering

functions. Units up to 5 kV·A capacity operating on single phase are normally configured for direct connection to final circuit distribution, although for security purposes it is likely that dedicated upstream circuits would be used. Many units employ integral 13-A socket outlets with sufficient intelligence to shutdown loads gracefully as the end of battery autonomy approaches.

'Off-line' systems are constructed with similar components to 'on-line' units, the fundamental difference being that, during the period that the mains is within acceptable tolerances, the load is served directly using various filtering and surge suppression techniques. The rectifier, battery and inverter limb are only employed when the input conditions are unacceptable or the mains fails. Different methods of connection of the battery-fed supply are employed varying from high speed circuit breakers to static switches with transfer times as short as one millisecond.

The other option available in this range, but limited to single phase operation, is the ferro-resonant transformer. In these systems the transformer has a dual primary input, one being from the mains the other from a conventional rectifier, battery and inverter leg which operates only when a mains interruption is detected. The few microseconds taken by the inverter to come on-line is covered by the stored energy in the transformer. From the user's viewpoint, the issues that separate the two types of system come down to one of risk: the off-line system is known for its 'blips' on transfer and, unless the inverter is constantly monitored, the likelihood of failure when called to act 'in anger' is a constant worry. Transient let-through is not a major issue at these kV·A ratings, if line conditioning and suppression is fitted for 'clean power'.

9.2.2 Free standing units up to 160 kV·A

The capacity level of 160 kV·A selected for this group is approximate, as are all thresholds used within this section of this Guide.

Single phase units are available up to approximately 50 kV·A, but beyond this level the systems tend to have three-phase inputs which are preferable when considering supply distribution arrangements. The 160 kV·A figure is significant when considering losses as it is probably the highest level that would be acceptable for heat rejection into a computer suite or office area. A further consideration in this respect is the noise emission from units within an occupied working environment, which could exceed acceptable criteria when operating at a high load factor.

By far the greater number of units, if not all, in this category now use transistor technology in the inverter. The systems follow the conventional dual conversion arrangement of rectifier, DC limb and inverter with a static bypass limb connected to a common output via a static switch. A typical arrangement is shown in Figure 9.5.

The method of operation follows that of smaller 'in room' units with similar voltage and frequency tolerance levels. During periods when the equipment and the input supply is healthy the critical load is isolated from any mains disturbances by the DC stage. During equipment faults or overload conditions the load is transferred to the bypass

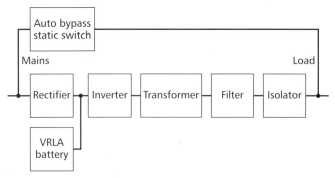

Figure 9.5 Typical transistor static UPS with auto-bypass

limb via the static switch and supported only by the unconditioned mains.

Refinements are available including full maintenance bypass arrangements to allow safe and easy servicing of equipment whilst feeding the critical load from raw mains. Remote alarm, diagnostics and metering facilities are optional features allowing interfaces with building management systems (BMS), tailored remote monitors, or even the load, via (typically) RS232 ports. In general, however, due to the ever increasing range of BMS equipment (each having its own programming and communication protocol) it is usual to have a volt-free contact interface to the BMS. The majority of UPS cubicles incorporate a standard mimic diagram and metering, giving input and output data. In some cases, at the lower kV·A ratings, final circuit protection in the form of miniature circuit breakers are also integrated to obviate the need for additional PDUs. All units incorporate main output circuit breaker protection with facilities for isolation. The majority of units in this range incorporate integral batteries within the module up to approximately 30 kV·A but, above this, additional stands, cladded racks or cabinets are required for either local or remote location.

In this range of equipment the use of thyristor technology in the inverter section is almost non-existent although such systems are still in operation.

9.2.3 Systems of 160 kV·A and above

From 160 kV·A upwards the technology used varies greatly. The selection and analysis of units becomes more difficult and time consuming and will be governed by a number of aspects discussed elsewhere in this Guide. All units from this capacity upwards use forced ventilation methods within the cabinets but heat emission on this level would be unacceptable in computer rooms or office environments. The majority of the equipment is therefore more suitable for plant room environments. This becomes more pertinent when considering large capacity parallel systems and is obviously imperative where diesel systems are proposed.

The technologies used in this range generally include thyristor inverters, transistorised inverters and rotary equipment. The rotary category can be further divided into those using batteries as a method of stored energy and units using diesel engines to maintain the output during periods of mains failure. Diesel supported units fall into two types, the first relying upon kinetic energy to bridge the gap between mains disturbance and a rapid diesel engine start and the second using batteries for short term

energy storage, the diesel engine only starting up when an extended mains failure is experienced, see section 9.2.7.

Where standby generation is not installed then the battery autonomy time has to be selected to cover the orderly shutdown of the load, as the period of mains failure is unlikely to be predictable. Also, the UPS is unlikely to support the cooling equipment necessary to overcome heat gains both from the UPS and equipment and an early shutdown would be necessary to prevent excessively high temperatures.

9.2.4 Static systems with thyristor inverter

It is unlikely that users of this document will need to consider the performance of thyristor inverters but, for completeness, they are described briefly here.

In common with other forms of static system, these systems comprise a controlled rectifier, a battery supply from the DC link and an inverter using a thyristor switching device to produce a stepped square wave output which includes a high level of unwanted harmonics. To improve the output waveform it is necessary to install filter circuits comprising inductors and capacitors which effectively increase the impedance to the load and can increase total harmonic voltage distortion (THVD) on the output.

In the normal operating mode (where the mains input to the UPS system is healthy) the DC link is held at a voltage sufficient to maintain the float charge to the batteries, which are generally permanently connected. When the mains voltage is outside the normal operating tolerances or the supply is lost, the inverter power is drawn from the batteries until such time as their terminal voltage reaches a preset cut-off level. At this time it is usual for an alarm to be raised prior to disconnection of the load but supplies should have been restored, by supporting generators, allowing the system to revert to normal operations. Upon restoration of the input supply the rectifier will simultaneously recharge the batteries and power the inverter.

The forced commutation of the thyristors involves power capacitors, one of the least reliable of the electronic components and which reduce the potential MTBF of the system. In addition the reflected harmonics onto the DC bus (across the battery) are high and thought to be detrimental to the service life of valve regulated lead-acid batteries, if not fully damped by filters.

Thyristor inverter systems must be rated for the peak, not RMS, current demand of the load and are available in single module form up to a maximum of 800 kV·A and, by parallel operation, can offer output capacities of many MV·A.

9.2.5 Static systems with transistor inverters

Transistorised systems comprise the same basic components, the transistor technology being used to switch the DC to produce an AC waveform on the output of the UPS.

Historically, there has been a limit on the switching capabilities of transistors caused by the semiconductor materials and for this reason the largest modules have generally remained below 400 kV·A capacity. This relatively high load handling capability has been achieved primarily by using parallel transistor paths and complex control techniques for larger size modules. This system uses the high frequency switching ability within the transistor to build up a chain of pulses, the width of which can be modulated to produce maximum power to coincide with maximum current demand of the load. These width-modulated pulses are fed into the primary side of a transformer specially designed to produce a sinusoidal waveform on the secondary side to which the load is connected. The fast response of the system enables non-linear loads to be fed more easily as the pulse width can be adjusted to support peaks in a distorted current waveform without undue voltage collapse. The regulation of the inverter is by a high frequency reference waveform with output voltage monitoring fed back directly to the inverter to provide rapid response to the load characteristics.

An inverter of this type has a limited overload capacity therefore the system must be arranged to transfer to mains upon a downstream fault, as long as the mains is present and within tolerance. The rating of the inverter must take into account the peak current demand of the load, not the RMS value, which is why most manufacturers limit the ratings to a maximum crest factor of 2.5–3. A typical transistor inverter, output transformer and filter arrangement is shown in Figure 9.6. To achieve a system design, using such UPS types, which can cope with a higher crest factor it is necessary to derate the UPS module. Output filters, with transistor switching operating in the 1 kHz to 20 kHz band, tend to resonate at high frequencies and offer limited assistance when dealing with harmonics.

The complexity of the transistor inverter, often using up to 48 switching devices, offsets to some extent the advantage of not having forced commutation capacitors as used in a thyristor inverter. General operation of the system and response to voltage and frequency deviations on the input are similar to that discussed for thyristor systems, but much faster in response to a load step.

9.2.6 Battery supported rotary systems

Battery supported rotary UPS systems are available in various forms with individual module ratings of <150 kV·A up to approximately 1700 kV·A. These units

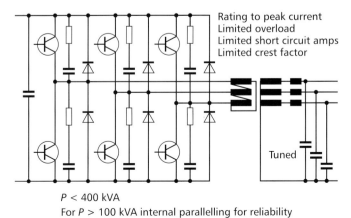

P < 400 kVA
For P > 100 kVA internal paralleling for reliability

Figure 9.6 Transistor inverter, output transformer and filter arrangement

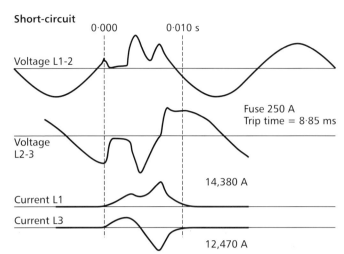

Figure 9.7 Voltage and current waveforms for 625 kV·A series on-line UPS

can be operated in parallel to unlimited combined output capacities. The output waveform is sinusoidal and rotary systems can produce very high short-term fault currents for clearance of downstream protective devices. Series on-line rotary systems do not use the bypass and mains supply to clear downstream faults. Sub-transient reactances as low as 5–6% result in 15 times the nominal current being generated to clear such faults, without the mains being available. All faults must be cleared within 10 ms to ensure other consumers retain power supply. Figure 9.7 shows the voltage and current waveforms of a 625 kV·A series on-line rotary module clearing a 250 A fuse in less than 10 ms. Rotary systems of this low impedance type can handle load crest factors up to 15 without de-rating, and need only be sized for the RMS value of the load current.

A derivative in common use is a rotary converter which is a synchronous machine in which the motor and generator windings are combined in an armature and rotor. The rotor is DC-excited and incorporates a special harmonic damper winding, similar to the squirrel cage on an asynchronous motor. The AC flux rotates carrying the rotor with it and by separate DC excitation of the rotor the output voltage of the generator windings is maintained at a constant level. Energy transmission to the rotor is brushless and any harmonics in the motor or generator windings are largely suppressed by the damper winding. This particular converter is fed by a conventional arrangement of rectifier, DC limb and inverter system. The inverter in this type of system is commutated by the terminal voltage of the synchronous machine. Therefore it has no forced commutation capacitors, consisting only of six power thyristors, which has proven to be highly reliable. Figure 9.8 shows the normal configuration, with the load isolated from the inverter by the brushless machine and with the complete absence of capacitor filters. The system is augmented with internal redundancy by the introduction of a thyristor switch leg which operates around the rectifier/inverter limb.

Figure 9.9 shows a block diagram of this redundant system in which, whilst the mains is within tolerance, all the power for the load flows through the static switch path into the motor winding of the combined machine. This provides good efficiency and the manufacturers claim an MTBF for this system of over 630 000 hours without recourse to the automatic and no-break mains bypass (not shown in the diagram). With the mains bypass and

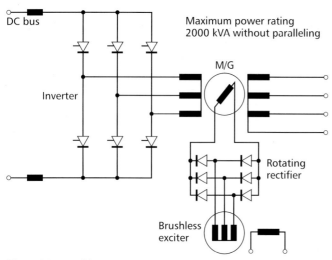

Figure 9.8 UPS with rotary converter

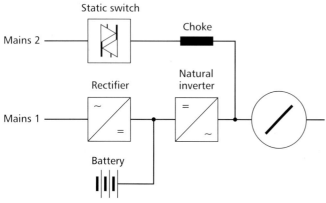

Figure 9.9 Internally redundant series on-line rotary UPS

average quality mains supply the MTBF rises to seven times that of a transistor static UPS.

A further system is available which introduces the thyristor switch leg on the output side of the machine (in parallel on-line mode), offering greater efficiency but this is not generally recommended for computer, data processing or communication loads that are susceptible to mains disturbances, or where the load harmonics are to be prevented from reaching the incoming mains supply. Parallel on-line systems are susceptible to mains frequency deviations.

9.2.7   Diesel supported rotary systems

Ratings for systems of this type generally range from 100 kV·A to approximately 2000 kV·A per module with the option of paralleled systems up to approximately 10 MV·A. There are two types of diesel rotary systems:

— kinetic energy storage (2–4 seconds)

— battery energy storage (normally 5 minutes).

9.2.7.1   Kinetic energy systems

One of the most common kinetic energy systems currently available comprises a diesel engine with free-wheel clutch, an induction coupling and a synchronous generator coupled to a split coil choke. In normal mode the electrical energy is supplied without conversion from the mains, the emergency energy source being connected in parallel and supplying reactive power to improve input power factor.

The alternator acts as a motor in the normal mode and is solidly coupled to the outer windings of the induction coupling rotating at 1500 rev/min. By excitation of this outer winding the inner part reaches a speed of 1500 rev/min in relation to the outer, or 3000 rev/min absolute. The inner winding is separated from the diesel by the free-wheel clutch. On mains failure or unacceptable deviation, the mains input contactor is opened and the diesel commanded to start. At the same time the DC winding of the induction coupling is excited and acts as an energy source and the outer winding of the coupling drives the AC machine, which acts as a generator rather than a motor. The diesel is required to start and achieve rated speed in 2–3 seconds upon which the clutch is closed. Excitation of the induction coupling is varied to regulate the output voltage and frequency. On mains resumption the system is regulated to parallel with the mains whereupon the clutch disconnects the diesel and the inner part of the induction coupling accelerates to its 3000 rev/min operation. The diesel is run for a short time (typically 5 minutes) off load, in order to cool and in case of further disturbances.

In areas of poor mains quality the diesel can be required to start several times per day. Black smoke emissions, from the rapid start, have to be taken into account.

9.2.7.2   Battery-backed systems

The battery-backed system uses a different configuration of similar components. The rectifier/inverter limb is used to provide the DC source, whereby a battery can be charged and connected to provide short-term power, typically five minutes, under mains failure conditions, thus driving the motor/generator from the output of the inverter. With this mode of operation using the batteries for short term support the diesel can be commanded to start off-load and with a ramped (and hence low smoke) acceleration characteristic and controlled to synchronise with the motor/generator shaft thus allowing minimal stress connection to the shaft via a clutch arrangement. With the machine operating in a stable manner on the diesel engine other loads can be added to the motor winding terminals (the machine now operating as an alternator) thereby offering the dual facilities of essential power (with a short break) and uninterrupted (UPS) power from the same system, but isolated from each other. Power is only drawn from the mains once the reliability of voltage and frequency has been restored. The environmental advantages of the battery backed diesel system are clear since the diesel engine is only started after a confirmed mains outage, rather than short duration deviations. The capital costs and space requirements of the battery installation are offset by the minimal engine service requirements, but the battery replacement costs equate to the refurbishment of the diesel engine in the other system.

Other kinetic energy systems are available using a single motor/generator shaft speed but using the system inertia to 'bump start' the diesel engine through a clutch held-off by the mains. In order that this starting procedure can be achieved whilst maintaining output supply tolerance a large amount of kinetic energy needs to be available. The transfer creates stress in the system and both safety and containment need to be considered.

It should be noted that systems operating in parallel to the mains do not provide electrical (galvanic) isolation. Harmonics can therefore be passed in both directions.

9.3   System selection

9.3.1   General

UPS systems are, with only a few exceptions, applied to critical loads operating on a 24-hour basis, 365 days per year, and which depend on both the quality and continuity of the power supply. At the same time as the need for UPS systems has expanded, the nature of the load equipment has changed, with switched mode power supplies drawing non-linear current. High concentrations of data processing loads in small areas can reflect high levels of harmonic currents onto supply systems with consequential effects on all elements feeding the load.

It is therefore important that, prior to the selection of UPS equipment, the load is established in its fundamental form and investigations carried out to establish the extent of non-linear content, the 'step' loads that are likely to be encountered, and the degree of phase imbalance that can be expected. Figure 9.10 shows an in-rush current for a modern high power processor. Due to the high investments made on larger systems, future expansion should be considered in order that either extra capacity is bought initially or provision made for easy expansion. These investigations should include supply monitoring of existing installations where UPS is being introduced or, on new projects, the monitoring of similar installations to those proposed. In both cases load profiles should be established.

Figure 9.11 shows a measured load current waveform in a large data centre of a major bank. The harmonic content, by order and total, is shown. The peak current is 1350 A. Increasingly, large UPS systems are being installed not to support central computers but large numbers of personal computers, terminals and file servers.

For example, an installation comprising 400 PC terminals and support equipment requires a UPS rating in excess of 400 kV·A and, importantly, at a current total harmonic distortion (THD) of 110% with crest factor >5. Failure to take these factors into account would result in an unreliable UPS solution and very high rectification costs.

It is recognised that equipment has to be selected on cost, size and ease of installation but these should not be the main criteria for UPS systems. Where non-linear loads are expected, the main criteria should be the way that the UPS equipment will react and the quality of conditioned power it will supply to different loads. Many manufacturers' performance specifications will only relate to operation on a linear load and a simple reference to crest factor will not resolve difficulties.

Other criteria that will have a bearing, particularly upon larger system selection, is the site location, its affect on adjacent occupiers, the frequency and length of local supply interruptions or departures from acceptable voltage and frequency limits, and the availability of emergency or secondary power supplies. The electrical distribution arrangements will also become an issue in selection with respect to the harmonic current produced by the input arrangement of the UPS system, and the effects these harmonics would have on other load groups within the distribution system. If the elimination of input harmonics are a critical matter, and the load is sufficiently large, then a series on-line rotary system should be considered. Without filters, this type of equipment achieves a THD of <2%, regardless of load current THD. The specifier should be aware of input current THD quoted in manufacturers' specifications since they generally apply to linear loads — a situation rarely found in practice.

9.3.2   System configuration, reliability, availability and MTBF

9.3.2.1   Configuration

There are three basic UPS module configurations to:

—   provide for the initial load

—   dictate capacity expansion steps

—   provide an MTBF to suit the user requirements.

These configurations are, in ascending order of reliability:

(*a*)   parallel power modules

(*b*)   single module

(*c*)   parallel redundant modules.

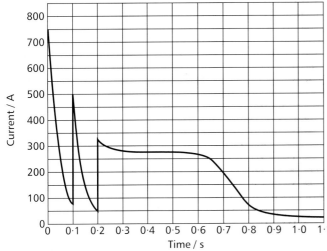

Figure 9.10 In-rush current for a high power processor

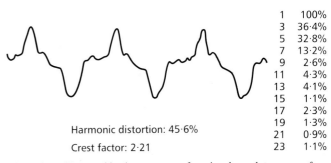

1	100%
3	36·4%
5	32·8%
7	13·2%
9	2·6%
11	4·3%
13	4·1%
15	1·1%
17	2·3%
19	1·3%
21	0·9%
23	1·1%

Harmonic distortion: 45·6%

Crest factor: 2·21

Figure 9.11 Measured load current waveform in a large data centre of a bank

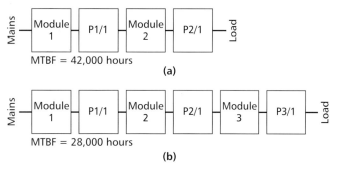

MTBF = 42,000 hours

(a)

MTBF = 28,000 hours

(b)

Figure 9.12 Reliability block diagrams for (a) two-module parallel and (b) three-module parallel configurations; assumes MTBF of paralleling control = 500 000 h and MTBF for each module = 100 000 h

Parallel power

The load capacity is met by adding together modules of smaller capacity, e.g. a 500 kV·A load is supplied by 2 × 250 kV·A modules. Obviously the reliability is the low since a fault in either module will cause the system to trip to bypass (the remaining module(s) being overloaded) and the critical load being exposed to raw mains, if the mains supply is present or within tolerance when the fault occurs. If the mains is not within the specified limits, the load will be lost.

The future expansion steps are 50% of the initial load but multi-module (>2) configurations become increasingly less reliable and are rare in practice. Figure 9.12 shows the reliability block diagrams for two- and three-module parallel power configurations, which clearly shows the reduction in theoretical reliability.

Single module

Single module systems offer an MTBF of the selected UPS module technology, but will vary greatly between systems. Indeed, single module, internally redundant, rotary systems have MTBF figures stated as higher than redundant static UPS systems when compared on the same basis. System capacity expansion is in steps of 100% of the initial load.

Parallel redundant

Parallel redundant system configuration can be (1+1) or (n+1), so that for a 400 kV·A initial system load the possible solutions would be 2 × 400 kV·A or 3 × 200 kV·A or 4 × 130 kV·A etc., having one full module running in excess of the system load. In the event of one module failing or being switched out of the system for maintenance, the remaining module(s) can deliver the system load. Each of the two options, (1+1) or (n+1), has advantages and disadvantages:

— (1+1) is theoretically the most reliable configuration since if one of the modules is switched off the other module handles the load without being in parallel power connection. It is the most compact multi-module arrangement. Each of the modules run at only 50% capacity so the 50% load efficiency figure is important. Also the module's ability to handle a load step of 50% (i.e. between 50% and full load) has to be proven. System

expansion steps must be in 100% steps of the initial system load.

— (n+1) is less reliable than (1+1). Apart from having more modules (and therefore more risk of one failing) when the redundant module is switched off by failure or for maintenance then the remaining sets run in parallel power. The first fault or overload will trip the system to bypass onto the raw mains (if present and in tolerance at that moment). Obviously one advantage is that each of the modules runs at a higher percentage load and therefore, probably, at a higher efficiency. System capacity expansion steps are 50% (2+1) or 33% (3+1) etc.

9.3.2.2 Reliability

All UPS manufacturers express the reliability of their modules in terms of MTFB and it can be shown that these figures are based upon a mixture of gathered data, practical experience and reference to component reliability standard data. In addition, it is a fact that a module MTFB is strongly influenced by the quality of the mains supply, the battery, the availability of service and whether or not the manufacturer has included the auto-bypass and availability of the mains in the calculation. The system designer must therefore question the basis of MTFB and make comparisons between differing data. Certainly the two most important features are the quality of the design/manufacture of the UPS modules and the availability of spare parts, 24-hour service support, without extended travelling times.

Having established the initial load (kV·A, power factor, THD of load current and crest factor) the specifier has to determine with the client/end-user the anticipated growth (steps and time scale). Armed with these facts, the question of system integrity must be reviewed. A system design which regularly exposes the load to raw mains supply cannot be considered an adequate solution for an application requiring continuity of conditioned power. The nature of the user's business will largely dictate what level of security of supply will be required. For example, a small computer centre carrying out batch processing tasks will have to expend only a few person-hours to recover from a system crash. However, the data centre for a large bank will have extensive on-line and remote terminals, including automated cash machines, and be processing data on-line, 24 hours per day. It could take 6–10 hours simply to restart the system after a power supply outage, with full recovery taking considerably longer, resulting in a major financial and administrative disaster to the organisation.

9.3.2.3 Availability

The term 'availability' has been used as a specification criterion for the power supply to the critical load. This is not a valid criterion since the user will be interested in long term reliability, not a statistical value, as follows:

$$\text{Availability} = \text{MTBF} / (\text{MTBF} + \text{MDT}) \qquad (9.1)$$

where MTBF is the mean time between failures and MDT is the mean downtime, including travel time.

However, note that:

— for MTBF = 10 years and MDT = 1 hour:
   availability = 0.9999885

— for MTBF = 1 month and MDT = 30 seconds:
   availability = 0.9999885

— for MTBF = 1 day and MDT = 1 second; availability
   = 0.9999884.

This shows that the availability is similar for the three situations, even though a 1 second loss of supply per day would be totally unacceptable in the case of a computer installation for which the start-up time was six hours.

In this case:

— for MTBF = 10 years and MDT = (6 + 1) hours:
   availability = 0.99992

— for MTBF = 1 month and MDT = 6 hours + 30 secs:
   availability = 0.9917

— for MTBF = 1 day and MDT = 6 hours + 1 second;
   availability = 0.8.

9.3.2.4   Mean time between failures (MTBF)

It is incorrect to confuse MTBF with a predicted time to failure for a single module or, indeed, a system. The theory behind MTBF calculations shows that at the end of the calculated MTBF period only 0.37 of the measured sample group is still in existence.

What MTBF can be used for is comparing systems. (Often MTBF is the only data available other than intuitive decision making.) With the same basis for the MTBF calculation and similar numerical techniques then MTBF figures can be relied upon to give a fair comparison of system integrity, but not a prediction of uninterrupted service time.

The MTBF figure is the mean time between events which result in the load being exposed to raw mains and does not include maintenance shutdowns or operator error.

Different manufacturers will claim varying MTBF figures for their products, with most of the differences arising from assumptions made about mains quality, whether or not the bypass circuit and raw mains supply to the load is included, and the travel time assumed in the mean time to respond (MTTR).

Typical MTBF values for on-line modules, not including the bypass circuit, are given in Table 9.2.

Table 9.2 Typical mean times between failure (MTBF) for on-line modules

Module type	Mean time between failures (MTBF) / h
Thyristor inverter, static (reference only)	35 000 to 45 000
Transistor inverter:	
— static	75 000 to 100 000
— rotary	100 000 to 170 000
Rotary with internal redundancy	610 000 to 700 000
Diesel rotary:	
— kinetic energy	210 000 to 350 000
— battery backed	800 000 to 1 000 000

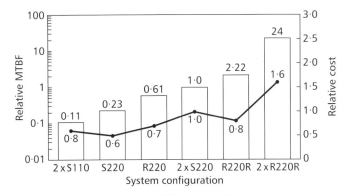

Figure 9.13 Mean time between failures (MTBF) for the most common configurations

The question of MTBF for a multi-module system (whether parallel power or redundant) is further complicated if the design/manufacturer chosen has equipment which has a single paralleling point and paralleling control system.

To enable a fair comparison between system MTBFs, it is necessary to include the auto-bypass in the calculations since static systems rely on the bypass to clear faults. Figure 9.13 shows the relative MTBF of the major options, taking the 'unity' MTBF as a parallel redundant (1+1) transistorised static series on-line system.

The configurations shown are;

— 2 × S110: parallel power pair of 110 kV·A static modules

— S220: static module rated at 220 kV·A

— R220: rotary module without internal redundancy

— 2 × S220: redundant pair of 220 kV·A static modules

— R220R: rotary module with internal redundancy

— 2 × R220R: redundant pair of 220 kV·A rotary modules.

Obviously the specifier has to address the equation of capital cost, energy consumption, plant space and required resilience over the life of the installation. It must be repeated that MTBF is not a prediction of equipment life but can be used to make valid comparisons of resilience and reliability of the critical power supply.

9.3.3   Dedicated range up to 5 kV·A

The selection of this range of UPS equipment should be simplified by the limited equipment that is usually connected and, for capacities up to 5 kV·A, the equipment will generally be a standard purchase item selected on a 'preferred manufacturer' basis. Load analysis will be easier but consideration should be given to rapidly changing equipment demands. Processors in modern personal computers have a current total harmonic distortion (THD) of ~70% and crest factors in the order of 8, but the current is so small in relation to the smaller UPSs that de-rating occurs by accident rather than by design.

Initial investigations should revolve around the susceptibility of the load equipment to voltage and frequency disturbances, noise spikes and 'micro' interruptions. If none of these are of major concern then an off-line UPS will provide continuous power for the specified period during major supply interruptions. Should any of these criteria cause problems to the load then on-line or the ferro-resonant solution should be investigated.

Having established what equipment will be served by the UPS, it is important that the ratio of peak-to-RMS current (i.e. crest factor) of the load is supplied to the manufacturer. As stated above, at this size the equipment would normally be a standard purchase item but if engineering analysis is critical various factors need to be considered, including the following:

— maximum output power with linear load at a power factor of 0.8

— maximum output power, non-linear load at a given crest factor

— equipment losses including battery charging

— maximum input current (per phase if necessary)

— part load efficiencies at 25%, 50% and 75% load

— type of rectifier used, e.g. 6-pulse, 12-pulse etc.

— dB(A) at 1 metre with 100% non-linear load

— compliance with the EC Directive on electromagnetic compatibility[1]

— MTBF of equipment or system configuration

— input current control on power restoration

— input voltage limits

— input frequency limits

— output voltage regulation at steady state, 50% load step and 100% load step

— transient recovery time

— THVD of output with non-linear load for all harmonic orders and max single order

— harmonics reflected by the UPS into the mains

— output frequency stability

— frequency synchronism range

— acceptable temperature and humidity conditions.

Having analysed the responses to the questions listed above, other considerations will revolve around size, weight, aesthetic appearance, acoustic control (if installed in an office environment) and cost acceptability.

Ease of installation and the output load connection facilities will be a consideration and it is important that having made the investment in UPS that if a component failure were to occur in the normal power limb that a bypass arrangement is available to connect the load to raw mains without a break.

The battery should be sized to suit the supply infrastructure serving the input connection to the UPS to allow ordered shutdown of the load if necessary. Typically 10 minute batteries are provided but 30 minute battery packs are available and may be advisable if the UPS does not have the additional support of emergency generation or if large deviations in supply voltage and frequency commonly occur.

9.3.4 Free standing units up to 160 kV·A

These units are generally located within a data processing or communications area and provide support to either a single power distribution unit (PDU) or a number of them. As such, load analysis is more difficult to achieve and may involve discussion with a number of hardware suppliers. As a minimum, consideration should be given to the provision of 25% spare capacity for future use over and above initial load performance levels assessed. This should not include any de-rating of capacity for non-linear loads.

For new installations the fundamental load requirements and prospective crest factor should be analysed. It is recommended that, if the load is available prior to procurement of the UPS, the feed be monitored such that a load harmonic spectrum can be reproduced with enquiry documentation. This will ensure that correct de-rating factors can be applied by the manufacturers.

Output filters, although necessary in a basic form on static systems to improve the output voltage waveform, should not be specified to handle particular harmonic frequencies. If tuned filters are specified, based upon an existing measured load, and the load changes in its characteristics (e.g. an updated main processor is installed) then the UPS may not be able to maintain the specified output. Active filters can assist this but they are generally biased to a particular frequency.

Comparison of the relative performance of each manufacturer's equipment should be based on the list of performance criteria produced for the dedicated range up to 5 kV·A. In addition to these items responses should also be gained on the following:

— peak current/phase and the time before reversion to bypass

— ability of the equipment to be paralleled for future expansion

— compliance with Electricity Association Engineering Recommendation G5/4[2] and the methods by which it is achieved.

Compliance with G5/4 becomes more difficult with large equipment. To obtain absolute assurance from manufacturers, supply system impedance values may need to be provided which will prove difficult, particularly on new projects. Absolute levels of harmonic current are given for various stages of compliance using normal supply voltage references for points of common coupling (PCC). It is not uncommon (to prevent or limit the re-injection of harmonic currents into the mains or generator or other upstream distribution) that phase shifted 12-pulse rectifiers are necessary, rather than simple input filtering techniques. The merits of each are arguable and need to be individually assessed. Input filters have three disadvantages which need to be taken into account:

— decreased reliability of system (MTBF)

— increased running costs (lower efficiency)

— noise and heat output.

In addition, it is unusual that filters are fitted to the bypass. Load harmonics (often higher that the rectifier's harmonic distortion of the input current) are transferred to the mains supply side when on bypass. THD for the common forms of rectifier input current, at full load, are approximately as follows:

— 6-pulse: THCD ≈ 33%

— 12-pulse: THCD ≈ 11%

— 24-pulse: THCD ≈ 6%

Cabling techniques for systems up to 160 kV·A will inevitably use large conductors and the various manufacturers' equipment should be investigated to ensure that maintenance can be achieved by full use of integral bypass arrangements. If this is found not to be so, then consideration should be given to the installation of an interlocked fixed bypass switchboard.

Peak output current data are necessary to establish the maximum downstream protective device in the system and ensure fault clearance without reverting to the bypass limb. This may not be a problem whilst a mains supply is available to the bypass but if the mains had failed the results could be disastrous at the higher end of this range, and if cost is not a prime consideration. Rotary systems can reduce this potential problem by providing higher short circuit currents, without recourse to the mains.

When considering computer suite or office environments acoustic emission is a primary consideration and special treatment may be necessary to achieve acceptable levels. However after considering this and the other performance data indicated, selection will inevitably revolve around size, weight, ease of installation and, if appropriate, the appearance and noise level.

Manufacturers should be consulted on any methods of automatic fire extinguishing proposed to ensure that no damage can occur to vital components.

The UPS may be located deep into office space and, to ensure satisfactory support for the critical load, it may be necessary to consider additional remote monitoring/alarm. This will incur extra costs for most systems but will provide for a faster response by the operators. Alternatively, building management systems now have sufficient 'intelligence' to provide monitoring of output signals generated by the UPS on a simple points basis.

To minimise battery size and capacity and to ensure that shutdown in an orderly fashion is possible, consideration should be given to remote generator status indication. All parties should be advised of mains failure, generators 'on-load' or 'failed to start'. This should prevent any panic action that could occur in the event of a low battery voltage warning whilst still expecting emergency generator support.

9.3.5 Systems above 160 kV·A

As indicated in section 9.2.3, the technologies available on large systems are varied. Similarly the system configurations can be arranged to give almost ultimate security for the critical load. However these systems will often represent a large percentage of a building load and the possibility of reflected harmonics on the building distribution system will have a much greater bearing.

Historically, space requirements for rotary systems (either battery supported or diesel supported) were greater than those required for static systems, but technology has now advanced to reverse this situation. However, it should be acknowledged that diesel supported systems require additional service route requirements in the form of large input ducts to provide cooling and aspiration air for the engines. Exhaust routes are also required for the removal the combustion gases from the engine(s). This is relatively easy to achieve (with the necessary noise attenuation) at ground floor level and above but, if the units are installed below ground floor level, it can result in considerable loss of space on other floors.

Oil storage space for diesel rotary UPS can be compared with the battery space for static and battery supported rotary systems and it should be realised that, provided sufficient oil storage can be achieved for the UPS, that the normal standby generator will not need to support the UPS load which will achieve a capacity saving.

All systems, static or rotary, have predetermined levels of input voltage and frequency variation that they can accept whilst supplying the load. Where these limits are exceeded systems generally revert to their emergency mode of operation utilising the batteries or the diesel source to feed the load. The site location, proximity of adjacent occupiers and the frequency of supply deviations could therefore play a vital role in the selection of equipment. Establishing the supply limits that would initiate the emergency operation, is vital, particularly for all triport and parallel on-line 'high efficiency' systems which do not have a wide input frequency tolerance.

Having decided which are the acceptable systems and modular configurations the manufacturers should respond to similar questionnaire items listed for smaller systems. These questions should be based around an initial enquiry load together with a prediction of crest factor, power factor and a typical harmonic spectrum. Static systems generally incorporate filters to 'trap' predominant harmonics on the output. However it is commonplace to apply additional output filters to achieve THVD levels of less than 5% with a specified load. This can make it slower to react to load changes, and will usually be tuned to particular operating spectra which could change, either on a daily basis or over longer periods. The reliability, efficiency and audible noise output are all adversely affected.

Parallel modules need to have a common single output busbar and it is therefore necessary to give consideration to an output panel comprising paralleling and bypass facilities. This will generally incorporate a bypass breaker capable of handling the full load system current. It is therefore essential that, having selected the system and configuration, future expansion is considered at initial purchase stage such that sufficient circuit breakers and current carrying capacity are included on the output busbar and system bypass limb. This will ensure continuous operation or only minimal shutdown time if additional units are introduced when the system is on-line. However in larger systems it is unlikely that the distribution system and transformers have sufficient capacity to simultaneously support both critical loads and load bank loads during test procedures.

Most, but not all, static multi-module parallel or parallel redundant systems use logic control to supervise the operation of the paralleling and common static bypass. These controls are programmed such that if the redundant module is removed (for maintenance etc.) then upon a subsequent module failure the system will revert to bypass irrespective of whether the load is within acceptable limits for the remaining modules. This is not generally the case for any system with distributed paralleling control and rotary modules in particular.

With higher power systems feeding more distributed loads throughout a building complex it is essential that users are aware of system status. This should include:

— input supply status (all modules)

— modules available

— bypass status

— emergency power operation (i.e. battery discharging)

— generators failed to start.

To monitor load growth and characteristics, the output switchboard could incorporate facilities for the connection of voltage probes and clamp-type current transformers. This should be possible without interrupting the supply and may be achieved simply by a protected socket outlet (voltage) and a section of isolated insulated busbars behind a lockable cover.

Final selection should be based upon the technical responses provided by the manufacturers, space and ventilation aspects and the effects of the UPS on the power distribution of the building with consideration to standby/emergency power plants.

9.4   Bypass arrangements

For maximum flexibility in multi-module systems each UPS module should be served from individual protective devices on the input switchgear and similarly the automatic bypass unit should be provided with an independent protected feed. The automatic bypass limb will be required to feed the critical load under failure of the UPS modules and will be the source of energy, albeit unconditioned, under major maintenance or reorganisation. It is essential therefore that consideration be given to the ultimate load of the system in order that the bypass circuit breaker is sufficiently rated and does not require future replacement with possible major disruption.

The modules and auto bypass operate with a high degree of control interlinks. To allow facilities for full system testing without detriment to the critical load, consideration should be given at an early stage to a full maintenance bypass. This should take the form of an identical circuit breaker to that provided for the auto bypass separately fed from the input switchgear and linked directly to the output switchgear. To prevent inadvertent operation the maintenance bypass should be interlocked with the auto-bypass to prevent closure of the former until the latter is closed.

9.5   Normal supply considerations

The effects of phase controlled rectifiers within the input of most UPS systems can greatly affect the electrical distribution to buildings.

Small dedicated UPS systems will not generally have sufficient demand to create many problems although they will still reflect harmonics onto their feeder circuits. In areas where there is a concentration of data processing equipment that is deemed not to need UPS support, these harmonics may well cause corruption detrimental to the systems. Location of small units may need special attention to ensure that problems are not caused through electromagnetic interference, although modules should be manufactured in accordance with the EC Directive on electromagnetic compatibility[1].

Larger systems will need to be considered more carefully, the first step being to relate capacities to Electricity Association Engineering Recommendation G5/4[2] for the point of common coupling with the supply company mains which is applicable to the building.

The majority of UPS systems, with the exception of the diesel rotary kinetic energy system, use either 6-pulse or 12-pulse rectifiers. These themselves reflect harmonics back into the feeder system with 5th and 7th harmonics being predominant with 6-pulse equipment and 11th and 13th being predominant with 12-pulse. To reduce these levels on 6-pulse rectifiers to a standard in compliance with Engineering Recommendation G5/4 it is usual to employ capacitive filters which will also have the effect of improving power factor. However, at low load, these can produce an unwanted leading power factor and it is common to hold the filters out of circuit until particular load levels are achieved. The additional heat losses should be added to the module losses.

When viewing the harmonic currents developed by comparable size UPS systems with both 6-pulse and 12-pulse rectifiers it can be seen that filters are more necessary with 6-pulse operation rather than 12-pulse. These filters can also become a 'sink' for other harmonic disturbances on the upstream distribution system and maintenance will have to be increased to monitor the condition of the filters and the upstream supply.

When UPS systems are to be installed on common low voltage distribution systems in smaller developments the electrical distribution should be investigated at length to determine other sources of disturbances or equipment groups that will be affected by the harmonics of the UPS itself.

On larger systems, where the UPS can be separated from other building load groups and possibly fed from a dedicated transformer, expected harmonic current levels should be given to the transformer manufacturer so that the necessary de-rating factors or transformer core modifications can be made.

9.6   System considerations

The supply to any centralised UPS system should be derived from a primary distribution centre within the building and connected using a dedicated cabling system.

The UPS system will have a finite period of autonomy under its stored energy operation. For long term resilience the system should be backed-up by an emergency power source such as an automatic mains-failure diesel generator system. Again the feed(s) to the UPS system from any emergency power source should clearly be defined and dedicated.

With the exception of continuous process equipment, UPS systems will generally be required to support computers and other data and communications systems that will not only generate harmonics themselves but will also be susceptible to supply system harmonics. It is therefore important that the supply system be designed to limit the effect of harmonic distortion upon its components such as switchgear, transformers and generators. It is also important to prevent the spread of harmonics from the UPS.

In the UK, the majority of building distribution systems operate at 415/240 V and buildings have until recently been sufficiently small or of sufficiently low demand that the whole of the electrical demand can be sourced on site from a small number of HV/LV transformers. As such, critical load sections may often be sourced from a common LV busbar system, which could feed HVAC systems, lifts or general lighting and power systems. Isolation of harmonic distortion from these systems and the UPS must be a primary consideration.

Where a number of small UPS systems are proposed to protect a network of personal computers then it may be necessary to consider a PDU for the area, if necessary, incorporating an isolating transformer to limit the spread of distortion through the supply system. Single units up to about 3 kV·A will probably not create any more distortion than the load they serve and, as such, can probably be supplied from the general distribution system local to the area.

Systems from 3 kV·A single phase up to 20 kV·A three-phase will generally be sourced from sub-main distribution systems. In these cases it is important to ensure that manufacturers incorporate filter traps on the input side to prevent the spread of dominant harmonic frequencies into the general distribution networks. Equipment of this size will generally be in single module form and may not have inherent bypass arrangements, but it is recommended that some form of bypass switch be provided to feed the load when maintenance is carried out. This should take the form of an interlocked or lockable switch that can only be closed once a responsible person has ascertained that the input and output of the UPS are in synchronism. It should be remembered that a UPS is installed because the load is critical; if the supply fails it is better for the load to see a raw mains connection than not have a supply at all.

Larger systems of either single or multi-module configurations will generally be served from the main incoming switchboard supply or, if the system is sufficiently large, from a dedicated substation. If possible, it should be established at the outset what ultimate capacity will be required; if not, the designer should advise the client of any limiting factors that would hinder future developments.

Systems having an overall nominal capacity of up to approximately 400 kV·A are most likely to be fed from a switchboard having other mixed loads which, if linear in electrical characteristic, will serve to dilute the effects of harmonic currents. If the other feeds are serving high technology loads or any loads with distorted characteristics then consideration should be given to the de-rating of all upstream components. This de-rating exercise would also be necessary if the UPS load formed a major percentage of any feeder distribution system.

Systems of 500 kV·A and above are likely to be served from a dedicated transformer substation and switchboard and it is essential that the transformer manufacturer is aware of the input characteristics of the UPS equipment for de-rating purposes.

Similarly, emergency generators, whether dedicated to the UPS or serving mixed loads, will need to be given special consideration. Additional capacity may be necessary or low reactance alternators employed to minimise the effects of harmonic distortion on the generator output stability.

The emergency generator should be analysed to determine its load handling capability for step loads as it is likely that the UPS will form a large percentage of the generator capacity. Load acceptance levels of 50–60% of alternator capacity should be a target and if possible the specification for the UPS should incorporate a requirement to 'walk-in' on power resumption, either by module control or by load ramping on the module inputs.

The whole of the distribution system, both to and from the UPS, should be provided with a neutral conductor at least equal to each phase conductor. In some cases it has been found necessary to oversize the neutral conductor due to the affects of 3rd order harmonics, typical when single phase loads are connected to the three-phase UPS output network.

Switchgear should, if possible, be arranged to serve the immediate needs of the system and to allow for any anticipated expansion or additional modules. The majority of systems derive their battery charging feeds from the DC limb but some have battery chargers fed separately from the mains. System bypass switches, whether automatic or manual, should incorporate capacity for anticipated expansion and in multi-module systems this feed could be substantial.

This could create the need for overrated busbars, incoming switches and feeder systems when considering system discrimination. Figure 9.14 shows a typical switchgear configuration for a large, high integrity, system with expansion capability.

Lastly the regional electricity companies (RECs) are becoming increasingly aware of the effects of UPS and other inverter type loads and are more stringently applying Electricity Association Engineering Recommendation G5/4[2]. This document is primarily intended to protect consumers from electrical noise and disturbance caused by other consumers. The limits are based on the point of

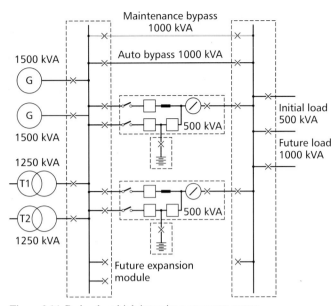

Figure 9.14 Redundant high integrity system UPS system

common coupling (PCC) to the network and give various stage limits according to capacities. Compliance should be requested as a UPS specification clause and acceptable limits agreed with the appropriate electricity supply company.

9.7   Site generated supplies

The relationship of emergency generators and UPS should be considered in detail. The effects of harmonic currents in the alternator windings, particularly when the UPS capacity is close to that of the alternator, can be problematic. To overcome this it is common to de-rate standby generators supporting UPS. A rule of thumb is a factor of 1.5, e.g. a 500 kV·A UPS fed by a dedicated 750 kV·A alternator. Low impedance or de-rated standard alternators and the method of excitation need to be considered. Particular attention should be paid to the possible effects of harmonics on the automatic voltage regulator.

UPS equipment of all triport (and most parallel on-line types) are susceptible to voltage and frequency disturbances on the input. If this is excessive the systems generally revert to battery mode or initiate starting procedures for generators. A diesel driven UPS will not normally be supported by standby generators but battery supported triport or parallel on-line systems will and, if frequent deviations in the feed are experienced, the system will constantly fall back to battery operation thereby reducing eventual battery autonomy in the event of a prolonged supply failure.

The 'power walk in' function is therefore of paramount importance to ensure that a large step load is not presented to the generators when they resume the input feed to UPS equipment thus causing excessive supply deviation.

9.8   Environmental conditions

Primary considerations relating to the environmental control revolve around the maintenance of room space temperatures in the range of 0–40 °C. Because of continuous operation of the equipment with inherent constant losses, control of the upper temperature limit is the fundamental concern. Lower limit control usually becomes a concern only during construction phases or in the case of major maintenance when temporary measures can be taken. Design ambient temperature, under full load operation, should therefore be in the range of 15–25 °C. Specific requirements for room space temperature, humidity and dust control should be discussed directly with all manufacturers considered.

Small dedicated micro-UPS units and, indeed, larger units of up to 100 kV·A associated with small computer or other critical loads will often have their environmental conditions controlled by the fact that they can be installed in the same room as the load that they are protecting. Critical considerations in respect of environmental control is the means by which temperature control is effected under emergency power conditions. Consideration may need to be given to split central plants providing reduced duty if emergency generator capacity is limited.

A further consideration is what would happen to the room temperature if the generator system does not start. The stored energy system will still feed the UPS equipment and its load for the period of autonomy. The system losses will be discharged into the area thus raising the temperature if cooling is not available.

In larger single and multi-module systems of any type, the primary calculation will be one of heat rejection. UPS systems provide continuous conditioned power to a critical load and are very rarely taken out of service. The problems are therefore security and resilience. If possible, the system should incorporate standby air conditioning plant and this should preferably be able to be maintained from outside the UPS room, thus alleviating the need for general maintenance staff to enter a secure area. For resilience, it should be possible to carry out full maintenance on air conditioning equipment, to change filters or maintain ventilation equipment without detriment to the full load operation in the UPS room.

Except for small systems, it is unlikely that these conditions can be maintained throughout the year purely by means of natural ventilation. It is generally recognised that, by the time thermal cut-outs in the UPS systems operate at their higher limit (i.e. ambient 40 °C but power components >100 °C), some damage may have already occurred to certain components which could reduce their life. It is also preferable to reduce the dirt intake into UPS rooms by the use of suitable filters with forced ventilation. To assist with dirt and dust control the room should operate at positive pressure and it is often easier to provide re-circulatory air conditioning to rooms for larger systems using minimum fresh air make-up and exhaust.

Fire control needs to be discussed both with the authorities and, because of the high investment and operational risk, with the client's insurers. It is often necessary to consider gaseous fire suppression systems with their attendant 'double knock' detection systems. UPS systems manufacturers should be advised of extinguishing proposals to ensure no damage to components or breaches of warranty. Further considerations are necessary on environmental controls such as automatic shutdown of plant, fire damper release and removal of gas.

All large systems create noise through air movement but highly distorted harmonic loads on static systems can create additional noise which will make adjacent areas uninhabitable unless acoustic treatment is applied. Manufacturers should be asked to provide operating noise spectra but this should be related to distorted loads. Diesel supported systems have unique problems of exhaust, jacket emission and cooling air noise, all of which are well known to the industry and will require special treatment, both to the air routes and the space. Smaller 'in-room' systems are generally treated at source and installed in attenuated enclosures.

The majority of UPS systems, of whatever size, incorporate forced cooling through the modules by means of a fan or fans. The cool air is sourced generally from the room environment by means of louvred entries in the enclosures. When designing it is important to ensure that these entries are not impeded and in larger multi-module systems where module heights can be as much as 2100 mm it is important that distribution of cool air is sufficient to ensure that modules are not starved or installed in 'hot spots'. Intake louvres are usually at low level with discharge at high level. If possible, consideration should be given to a suspended floor as a supply plenum, with interchangeable perforated tiles allowing adjustments to air patterns if necessary.

Integral fans are provided within modules because air passage through the units is essential for cooling purposes. Standby capacity has previously been suggested for room ventilation but UPS manufacturers should be questioned about the redundancy of their internal fan systems and the effect of single or multi-fan loss.

9.9 Battery/stored energy selection

The majority of UPS systems use lead-acid cells but nickel–cadmium cells are used in special cases. Types of batteries are considered in detail below, see sections 9.9.1 and 9.9.2.

Static systems and the most modern rotary systems generally connect the batteries directly across the DC limb of the system, the charging current being derived from the system input rectifier. In this way the batteries are permanently available should the input supply fail.

Rotary module designs of the previous generation incorporate a battery contactor and separate charger but needed a flywheel to provide enough time for the off-line battery to be switched into circuit. Modern designs have improved reliability by removing the battery contactor, and have reduced weight, cost and space requirements by not requiring a separate charger and flywheel.

Considering lead-acid cells as the most commonly used battery type there is one fundamental choice to be made: vented (wet) or valve regulated (sealed) pattern. Vented cells have a longer life, they are generally more expensive and require more space, which must be ventilated to prevent the build up of potentially explosive gases. Sealed cells are smaller, less expensive and require no special ventilation but have a substantially shorter life.

BS 6290[3] relates to batteries suitable for UPS duty and covers such points as the fundamental difference between 'design life' and 'service life'. The UPS supplier should be asked to confirm the design life and the anticipated service life of the cells. Batteries are usually considered to be at the end of their service life when 80% capacity remains. Batteries can be selected for 100% capacity at end of life by over-sizing the initial capacity by 25%, called an 'ageing factor'. This ageing factor does extend the actual life of the cells. In general two main factors affect battery life in UPS systems:

— AC ripple

— ambient temperature.

An extreme example, although historic, would be a thyristor inverter static UPS module (high reflected harmonics onto the DC bus) with insufficient DC smoothing, and batteries enclosed in cubicles with little cooling (i.e. ambient temperatures of 30–35 °C), the resultant service life could be as low as 2 years for a '10-year' battery.

Vented batteries (i.e. flooded cells of pasted plate, pasted rod or Planté designs, see below) are not so susceptible to temperature variation, because water loss can be corrected.

The ideal operating temperature for lead-acid batteries is 20 °C and it is therefore likely that some environmental treatment to battery rooms will be required. Vented batteries require an extract to fresh air and recirculation is not acceptable. The battery life is halved for every ten degrees above 20 °C. So that a 5-year design life cell operated at 35 °C (not uncommon in confined and poorly ventilated plant rooms) will last approximately 22 months in service, assuming no AC ripple across the battery.

Most manufacturers of valve regulated lead-acid (VRLA) cells now concur on the expected service life compared to the design life in UPS installations, with correct temperature and ripple control. Each manufacturer should be asked to provide details, but expect to be told that a 10-year design life product will provide a service life of about 8 years to 80% capacity. Preventive maintenance is essential, as is ensuring that the UPS supplier is fully involved in the specification, supply and installation of the battery system.

For UPS systems operating in computer suites or office environments valve regulated units are essential. Various types are available including gas recombination and gel electrolyte offering various life expectancies. Typical cells have a design life of 5 years or 10 years and, subject to cost limitations, the latter are preferable.

Vented cells have a life expectancy of 12–20 years, the upper figure applying to Planté cells. Unfortunately, because of their physical size and maintenance clearances, Planté cells can occupy a floor area of up to 4 times that of valve regulated cells when considering racks of similar height. Similarly their weight and cost can be twice that of valve regulated cells. Pasted rod (or plate) vented cells are available and offer a good compromise between size, weight, cost and service life. The choice is therefore between the low cost of VRLA cells and the longer service life, higher reliability, weight, and greater space/ventilation requirements of wet cells.

Selection of the autonomy period will be based upon the emergency supply arrangements for the building but will typically be in the 10–20 minutes time span. Batteries should be rated to the full load of each module at the prospective load power factor. Care should be taken in parallel redundant arrangements to ensure that batteries are rated for each module such that the full load and autonomy period can be achieved with (n) units, rather than $(n+1)$ units.

Battery racks should be provided with DC circuit breakers for isolation purposes and consideration give to shunt trip facilities for emergency shut down procedures.

For kinetic energy diesel rotary systems the long term stored energy will take the form of oil storage tank space and the capacity will be subject to both the tank location and the quality of the mains supply at the site. The storage capacity at full load consumption will probably equal the storage period selected for emergency generation but if frequent starting is expected due to large variations in supply voltage/frequency then some additional capacity may be necessary.

9.9.1   Battery types

This section relates to batteries in general and not specifically to those used in UPS systems.

To understand the principles of electrochemical cells and batteries it is useful to identify the reactions taking place in terms of chemical symbols. The terminology used in this section is given in Table 9.3.

9.9.1.1   Lead-acid cells

Lead-acid batteries are initially categorised according to the type of positive plate used in the cell, see Figure 9.15. They are further categorised according to the type of lead alloy used in the positive plate, and whether the cell has a vent, or is fitted with a valve.

Planté cells

The Planté positive plate is a thick sheet of lamellated pure lead onto which a very fine layer of active material has been deposited. The purpose of the lamellated surface is to generate a large surface area within the plates, which directly relates to the capacity of the plate. The plates have an overall thickness of 6–8 mm.

These cells have a long life but the energy density (W·h/litre) of the cell is low, except at very high rates of discharge. The construction of the Planté cell allows the

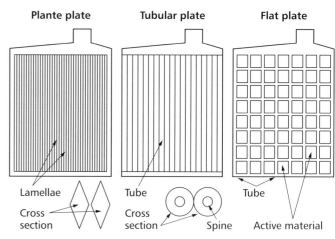

Figure 9.15 Lead-acid batteries: types of positive plate

cell to have a generous volume of electrolyte, which requires a low electrolyte specific gravity (1.220 kg/litre).

Tubular cells

The tubular positive plate is a fabric gauntlet of tubes containing active material and conducting lead–antimony spines. The antimony content varies between 0.5% and 8% and its purpose is to harden the lead. In comparison to the Planté cell the surface area is much higher in the tubular plate and this is reflected in a higher energy density, particularly at the lower rates of discharge. This higher energy density requires an electrolyte specific gravity of the order 1.240 kg/litre.

Pasted or flat plate cells

The pasted or flat positive plate comprises a lead alloy frame with an internal lattice which is filled with active material. The term 'flat plate' describes its appearance, and the word 'pasted' describes how it is manufactured.

Some flat plate cells are described as lead–antimony cells. This is because of the alloy used in the plate frame and lattice. The antimony content of the frame and lattice varies between 0.5% and 6.0%, though for long life and low maintenance periods the antimony level is kept as low as possible.

Other flat plate cells are lead–calcium cells and in such cases the frame and lattice alloy is hardened with up to 1% of calcium.

Pasted or flat plates achieve a similar surface area to that of tubular plates and therefore similar energy densities, again, particularly at the low discharge rates. This again

Table 9.3 Batteries: terminology and symbols

Item	Battery type	
	Lead acid	Nickel cadmium
Positive plate	Lead dioxide (PbO_2)	Nickel hydroxide ($Ni(OH)_2$)
Negative plate	Lead (Pb)	Cadmium (Cd) Cadmium hydroxide ($Cd(OH)_2$)
Electrolyte	Sulphuric acid (H_2SO_4)	Potassium hydroxide (KOH)
Gases/liquids produced	Hydrogen (H_2) Oxygen (O_2) Water (H_2O)	Hydrogen (H_2) Oxygen (O_2) Water (H_2O)

requires an electrolyte specific gravity of the order 1.240 kg/litre.

Vented cells

This is the traditional cell containing electrolyte in which the liquid level has to be kept between specified maximum and minimum levels. Each cell is fitted with a removable vent plug to allow gases to escape and to enable the electrolyte to be topped-up when necessary.

The types of lead-acid cells described above are generally vented cells.

Valve regulated cells

In valve regulated lead-acid (VRLA) cells, the electrolyte is immobilised between the plates, and gases are allowed to escape by means of a valve rather than vent plug. VRLA cells fall into two categories: those with electrolyte in the form of a gel, and those in which the electrolyte is immobilised by absorption into a glass fibre separator:

— *Gelled electrolyte cells*: cells of this type have been manufactured using tubular plates and pasted plates, in which the alloys used are either very low antimony alloy or lead–calcium alloy. Their behaviour is very similar to the vented equivalent except that the electrolyte used in the discharge reaction is limited to that contained directly between the plates. It is not possible to compensate for this loss of electrolyte by increasing its gravity, as this would ultimately increase float charge voltages to unacceptably high levels. As a consequence the specific gravity of the electrolyte used in gelled cells will be the same as for a vented cell but the energy density of the cell will be reduced.

   During operation, water will be lost from the cell as a consequence of gassing and this will allow oxygen recombination to take place. This reduces gassing and conserves water, but also reduces the voltage of the negative plate. This increases the float current and the internal heat generated by about 2.5 times.

— *Absorbed electrolyte*: cells of this type have only been manufactured by using pasted or flat plates. The frame and lattice alloys in these plates is either lead–calcium alloy or pure lead. The principal feature of the absorbed electrolyte cell that is different to other cells is the separator between the plates. These separators are usually made in microglass fibre which absorbs the electrolyte. The degree of electrolyte saturation in the separator determines the electrochemical and electrical behaviour of the cell.

   When the separator is totally saturated with electrolyte, the behaviour is identical to that of the gelled cell.

   When the saturation level of electrolyte in the separator is set to achieve maximum oxygen recombination (95%) within the cell, then the negative plate will have a lower float voltage. Under these circumstances, it is therefore possible to increase the electrolyte specific gravity without detriment to the float voltage, giving an increased

energy density for the cell. This type of cell typically operates with electrolyte gravities of 1.300 kg/litre.

9.9.1.2   Nickel–cadmium cells

Nickel–cadmium batteries are also initially categorised according to the types of plates used in the cell. In addition, there are small capacity sealed cells.

Pocket plate cells

The pocket plate cell consists of a plaque of perforated pockets filled with active materials. The plaque is pressed from thin perforated strip steel which has been annealed in a reducing atmosphere and, always in the case of the positive plate and sometimes in the negative plate, has been nickel-plated.

The electrolyte used is a nominal 20% solution of potassium hydroxide which has a specific gravity of 1.20 kg/litre but, as water is lost by electrolysis, cells are designed to allow the specific gravity of the electrolyte to rise to approximately 1.26 kg/litre when the electrolyte has fallen to its minimum level and water has to be added. When pocket plate cells are designed for use in arctic conditions the starting electrolyte is more dense, often with a specific gravity of 1.25 kg/litre which rises to approximately 1.30 kg/litre before water is added. The electrolyte concentration chosen is a compromise between cell life, conductance and freezing point. Lithium hydroxide is usually added to the electrolyte, typically in the range 8–20 g/litre LiOH in order to increase the life of positive plate. As electrolyte does not take part in the electrochemical reactions of the cell its discharge profile is very similar to that of the Planté cell shown in Figure 9.15.

Sintered plate cells

The sintered plate cell comprises a highly porous nickel substrate impregnated with the active materials.

The electrolyte used in sintered plate cells is essentially the same as for pocket plate cells but, in the sealed type, the specific gravity of the electrolyte is usually in range 1.25–1.30 kg/litre. The charge acceptance of the positive electrode, particularly at high temperatures, is improved by the addition of lithium[4]. However, the electrical performance obtained gives a superior value of energy density at all rates of discharge.

Pasted plate cells and hybrid cells

Various types of pasted plates have been made. These are a mixture of active materials in combination with plastic binders which are applied to a conductive metal mesh, dried and pressed to dimension. The negative electrode is the most successful and is finding application in combination with other positive plate types usually of the sintered type to form a hybrid cell. The electrolyte used is similar to that used in pocket plate cells.

Vented cells

The vented cell is the traditional flooded nickel–cadmium type, fitted with a vent plug which allows gas to escape and can be removed during topping-up. The vent cap is designed to minimise the ingress of air during topping-up, and thereby minimise carbonation of the electrolyte arising from the reaction with carbon dioxide. It often contains a porous disc to inhibit the progression of a flame front if the explosive gasses generated during charging should ignite.

Vented cells are manufactured using pocket plates, sintered plates, plastic bonded plates and combinations of these with pasted plates.

Sealed cells

Small capacity sealed cells have only been manufactured using sintered plates and negative bonded electrodes, but optimisation of the electrolyte volume and degree of separator saturation remain critical to the safe performance of these cells.

Oxygen recombination is therefore optimised to maximum efficiency allowing the cell to be effectively sealed. Whilst some designs intended for satellite operation are hermetically sealed, the majority have a pressure release valve or safety device which exhausts gas under fault conditions.

Because the oxygen recombination cycle is virtually 100% efficient, all the trickle charge current passing through the cell will generate heat which, however small, must be removed by ventilation.

9.9.2   Electrochemical reactions

9.9.2.1   Lead-acid batteries

The electrochemical reactions occurring in a lead-acid cell are summarised by the following equation:

$$PbO_2 + Pb + 2\,H_2SO_4 \underset{Charge}{\overset{Discharge}{\rightleftharpoons}} 2\,PbSO_4 + 2\,H_2O$$

The nominal voltage of the cell is 2 V. During the discharge reaction the positive plate (PbO_2) and the negative plate (Pb) react with the electrolyte (H_2SO_4) to produce electricity, lead sulphate ($PbSO_4$) and water (H_2O).

Because the electrolyte takes part in the reaction, the strength of the sulphuric acid decreases during discharge and increases during charge. The variation in electrolyte strength produces the following characteristics:

—   The internal resistance of the cell changes with the state of charge.

—   As the working voltage of the cell is related in part to the concentration of the electrolyte this will affect the voltage profile of the discharge curve.

—   At the end of discharge both plates are converted to lead sulphate, and if left in the dilute acid for any length of time, will not recharge efficiently.

On completion of the charge reaction, the cells will emit gas as the current flowing through the cell electrolyses the water. The reaction for this process is as follows:

$$2\,H_2O \underset{Electrolysis}{\overset{Gassing}{\longrightarrow}} 2\,H_2 + O_2$$

Water is consumed in the cell and it is for this reason that batteries need to be periodically topped-up with water.

There is, however, a variant of the lead-acid cell in which the vent plug is replaced by a valve, thereby preventing the battery from being topped-up with water. Some batteries of this type use the oxygen recombination principle to suppress the evolution of gas and therefore tend to generate a little more internal heat, in the absence of gassing, and have different float voltage characteristics.

9.9.2.2   Nickel–cadmium batteries

The overall electrochemical reaction taking place in a nickel–cadmium battery is usually given by the equation:

$$2\,NiOOH + Cd + H_2O \underset{Charge}{\overset{Discharge}{\rightleftharpoons}} 2\,Ni(OH)_2 + Cd(OH)_2$$

which is the sum of the reactions taking place at the two electrodes. These are usually shown as follows.

Positive electrode:

$$NiOOH + H_2O + e^- \underset{Charge}{\overset{Discharge}{\rightleftharpoons}} Ni(OH)_2 + OH$$

for which the standard electrode potential is usually given as $E^o = 0.49$ V.

Negative electrode:

$$Cd + 2\,OH^- \underset{Charge}{\overset{Discharge}{\rightleftharpoons}} Cd(OH)_2 + 2\,e^-$$

for which the standard electrode potential is usually given as $E^o = -0.809$ V.

Thus the theoretical open circuit voltage of the nickel–cadmium system is 1.299 V but this is an oversimplification since $Ni(OH)_2$ exists in two forms, known as α and β.

The true composition of the electrodes in the charged and discharged states is a matter of conjecture. The nominal open circuit voltage of the nickel–cadmium cell is usually taken as 1.20 V per cell.

The electrolyte does not take part in the electrochemical reaction as an active material, and therefore the characteristics of a Ni–Cd cell are different to those of a lead-acid cell:

—   The internal resistance changes very little during the discharge.

—   The cell is not sensitive to being left for long periods in the discharged condition.

On completion of the charge reaction, the cells will gas with the electrolysis of water in the electrolyte. The reaction for gassing and the need for topping-up are similar to those for lead-acid cells. However, a sealed nickel–cadmium cell has been available in small capacities (button cells and cylindrical cells) for over 30 years. These cells were the first to use the oxygen recombination principle. Whilst the cells are sealed, they have a pressure release valve or rupture disc to exhaust any pressurised gas under fault conditions. The sealed cells display different characteristics when compared to vented cells, with internal heat generation being higher and the applied system voltage levels being lower.

9.9.3 Mixed battery systems

Accidental mixing of the electrolytes of sulphuric acid and potassium hydroxide will not only destroy the cells but result in an explosive reaction.

It is recommended that lead-acid and Ni–Cd battery systems should not be installed in close proximity and maintenance facilities should be totally separate. Refer to BS 6133[5] and BS 6132[6].

9.9.4 Battery applications

Battery applications are given in Table 9.4. The reference code is used for identification purposes in subsequent tables.

9.9.5 Typical battery performance

Table 9.5 gives typical data on energy density, rate of discharge, capacity range, life and life cycles for various battery types. The information is based on laboratory tests at constant temperature for 'stationary' (STAT) applications (see Table 9.4).

9.9.6 Operating temperature

9.9.6.1 Effect on capacity

In all battery types capacity changes directly with temperature. Manufacturers will give the factors and methods by which allowance can be made for temperature.

9.9.6.2 Effect on battery life

The life of batteries decrease as the temperature increases by about half for every 10 K rise in temperature above the optimum for lead-acid batteries and every 20 K for Ni–Cd batteries. However, the converse is not true if temperatures fall below the optimum temperature. On the contrary, the float voltage should be increased in order that the battery can be maintained at the top of charge at lower temperatures.

9.9.6.3 Float voltage compensation at elevated temperatures

To conserve battery life, the float voltage should be reduced as the temperature increases. Appropriate values of compensation can be obtained from the battery manufacturers. Automatic float voltage compensation is sometimes provided in the charger control circuits.

Note: constant potential charging is not recommended for sealed nickel–cadmium cells due to the risk of thermal runaway. It should only be used where low rate charging is

Table 9.4 Battery applications

Application	Description	Code
Stationary	Batteries in a fixed location (i.e. not habitually moved from place to place) and which are permanently connected to the load and to the DC supply	STAT
Traction	Batteries used as power sources for electric traction vehicles or materials handling equipment	T
Starter	Batteries for starting, lighting and auxiliary equipment with internal combustion engined vehicles (e.g. cars, commercial and industrial vehicles)	SLI
Portable	Batteries of the valve regulated lead-acid or sealed Ni–Cd types for cyclic application in portable equipment, tools, toys etc.	P

Table 9.5 Battery performance data

Battery type	Capacity range / A·h	Energy density over stated discharge period				Typical lifetime (at 20 °C) / years	Typical cycle life (laboratory test) / cycles
		1 hour		15 minutes			
		W·h/kg	W·h/litre	W·h/kg	W·h/litre		
Vented lead-acid:							
— Planté	15–2000	8.43	18.06	4.25	9.11	20–25	400
— tubular	200–3500	10.54	29.4	5.07	13.64	10–15	1200
— flat plate	15–1000	11.02	25.3	5.31	12.65	3–12	400
Valve-regulated lead-acid:							
— gelled	200–1500	11.5	28.3	6.10	15.0	5–18	400
— absorbed	20–1000	15.48	43.6	9.83	27.7	8–13	250
Vented nickel–cadmium:							
— pocket	8–1540	15.8	27.2	8.2	16.27	<25	★
— sintered	11–440	21.7	32.53	13.96	20.93	<20	★
Sealed nickel–cadmium	15–160	28.47	81.27	23.61	67.4	<20	★

★ The absence of data reflects the absence of suitable laboratory life cycle test and not the performance of the product. Estimates of life cycle can be obtained from the manufacturers. Life cycle data is only important for some third world countries where the mains supply can be intermittent.

required and sophisticated control methods can be employed.

9.9.6.4 Storage life

Storage times for new, fully charged, batteries reduce with increase in temperature. The storage time reduces by a factor of two for every 10 K rise in the temperature. Values for shelf life can be obtained from the battery manufacturers.

9.9.7 System effects

In most applications battery performance will vary according to the system operating conditions.

9.9.7.1 Response mode operation

When the power supply fails the load is connected to the battery power source with or without interruption of power to the load.

In this mode the battery normal condition is that of receiving only a float charge provided by the charger. The arrangement is shown in Figure 9.16.

9.9.7.2 Parallel mode operation

The load is permanently connected to the charger and the battery so that when the power supply fails, the battery power source supports the load without interruption. The arrangement is shown in Figure 9.17.

Within parallel mode operation, the system (i.e. charger and the load) can react and have various effects upon the battery. The effects have been classified into groups of battery operational characteristics.

9.9.7.3 Standby operation

This is characterised by a permanent float charge condition being applied to the battery and a constant load in which the charger supplies the current to both the load and the battery.

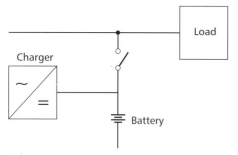

Figure 9.16 Arrangement for response mode operation

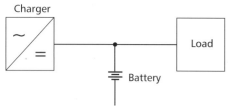

Figure 9.17 Arrangement for parallel mode operation

Typical standby float charge current profiles are shown in the Figure 9.18. The charger is able to supply all the current required by the battery and the load at any time.

Typical standby float charge current profiles with slight peak discharges are shown in Figure 9.19. In this case the charger is able to supply, on average, all the current to the load and the battery, including the recharge current. The battery is always kept in the fully charged condition provided the the value 'x' does not exceed the manufacturer's limits. A typical value for x would be in the range 0–5% of the average float current ($I_{f(ave)}$).

9.9.7.4 Battery buffer operation

This mode of operation is shown in Figure 9.20, and is characterised by a permanent float charge condition being applied to the battery and a variable load in which the charger supplies the current to both the battery and the load.

Whilst the average charger current meets the currents required by the load and battery, the battery cannot accept the positive cycle float current efficiently in the time available. In these conditions, the battery is not charged at all times, and this can lead to a 'walk down' in available capacity. To compensate for this situation there are the following alternatives:

— Operate the system at an elevated charge voltage and thereby keep the battery at the top of charge.

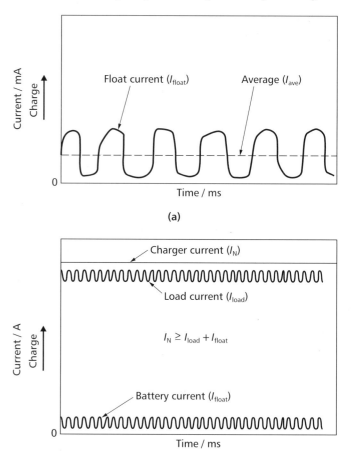

Figure 9.18 Battery standby operation; (a) float charge current with superimposed ripple current, (b) charger supplying full current to load and battery

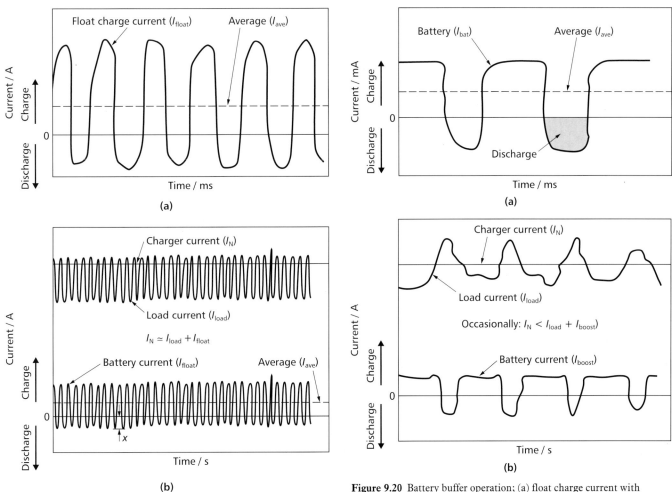

Figure 9.19 Battery standby operation with slight peak discharges; (a) float charge current with superimposed ripple current, (b) charger supplying full current to load and battery

Figure 9.20 Battery buffer operation; (a) float charge current with superimposed ripple current, (b) charger supplying full current to load and battery

— Operate the system at normal float charge voltages and give the battery regular off line charges at boost charge voltages.

— Operate the system at normal float charge voltages and allow the capacity to fall to its natural minimum which is usually approximately 80% of the catalogue value. This technique would require using a 25% larger battery capacity.

9.9.7.5 Battery shallow cycling operation

This is characterised by significant discharges of the battery together with a variable float charge condition. In these circumstances the service life would be determined by the number and depth of the discharge cycles.

Examples of this operation would be peak lopping systems or photovoltaic systems.

9.9.8 Battery type selection

9.9.8.1 Standards

Standards for lead-acid batteries are primarily organised according to application, while those for nickel–cadmium batteries are organised by type.

Table 9.6 Application and product standards for batteries

Application*	Code*	Application standard (lead-acid only)	Product standard	
			Lead-acid	Ni–Cd
Stationary	STAT	BS EN 60896: Part 1[8] BS EN 60896: Part 2[8]	BS 6290: Part 2[4] BS 6290: Part 3[4] BS 6290: Part 4[4]	IEC 60623[14] IEC 60622[15]
Traction	T	BS EN 60254[9] IEC 60254: Part 1[10]	BS EN 60254[9] IEC 60254: Part 2[10]	IEC 60623[14]
Starter	SLI	BS EN 60095: Part 1[11]	BS EN 60095: Part 4[11]	IEC 60623[14] IEC 60622[15]
Portable	P	BS EN 61056: Part 1[12]	IEC 61056-2[13]	BS EN 61150[16]

* See Table 9.4

Table 9.6 lists relevant application and product standards[3,7–15].

9.9.8.2 Battery applications and equipment specifications

Table 9.7 lists typical battery applications and related equipment specifications.

9.9.8.3 Selection method

The operational requirements that have to be determined are:

(a) service life

(b) cycling duty

(c) maintenance

(d) temperature range in service

(e) temperature range in storage.

The requirements against (a), (b) and (c) can be used with Table 9.5 to determine the type of battery most appropriate to the client's specification. Having established the type of battery and its application, the appropriate specification may be established from Tables 9.6 and 9.7.

The effects of (d) and (e) upon battery service life and storage life respectively will form part of the 'supplier enquiry procedure' for the batteries (see section 9.9.9.4), together with details of the benefits of float voltage compensation on service life, if necessary and available.

9.9.9 Charger selection

9.9.9.1 Load type and size

The selection of the charger is dependent upon the size and type of the load and the size of the battery. It is not the intention to provide the engineer with a guide on systems engineering, but to establish a position in which the appropriate questions can be asked of the supplier.

From the client specification the following operational requirements can be ascertained:

Table 9.7 Battery applications and equipment specifications

Battery application	Code*	Equipment specification
Central battery systems:		
— emergency lighting	STAT	
— UPS	STAT	BS 5266: Part 1[20]
— inverters	STAT	
— alarm systems	STAT	
— clock systems	STAT	BS 5839: Part 4[21]
Self-contained systems:		
— emergency lighting		
— UPS	STAT/P	BS4533-102.22[22]
— alarm systems	STAT/P	
— clock systems	STAT/P	
System free batteries (buildings):		
— starter	STAT/SLI	
— traction	T	

* See Table 9.4

(a) type of application (see Table 9.7)

(b) operational voltage range

(c) maximum power capability

(d) autonomy time of the battery (end of service life)

(e) battery recharge time and current.

9.9.9.2 Battery size

If the type of application is such that the load on the battery and the charger is not the load specified by the client, e.g. UPS or inverter applications, then either the whole of the specified requirement should be sub-contracted to the system supplier or an enquiry placed with them to establish values for (b) and (c) above to be supported by the battery and the charger.

The values of (b) to (e) pertaining to the battery will form a second part of the battery supplier enquiry procedure (see section 9.9.9.4) to establish the size of the battery concerned. Alternatively, for large, complex applications, the engineer may wish to calculate the battery size. For lead-acid batteries, a useful reference is ANSI/IEEE Standard 485[16].

9.9.9.3 Charger size

To establish the charger size the following operational requirements will be need to be known:

(a) maximum constant voltage

(b) maximum power

(c) maximum battery recharge current

(d) whether automatic temperature/float voltage compensation is provided.

The values of (d) form part of the charger supplier enquiry procedure, see 9.9.9.4 below.

9.9.9.4 Supplier enquiry procedures

Batteries

The enquiry procedure for batteries is in three parts as follows:

(a) Temperature effects upon service life and storage, and float voltage compensation benefits, see 9.9.8.3 items (d) and (e).

(b) Battery size, see 9.9.9.1 items (b) to (e) and 9.9.9.2.

(c) Battery dimensions, including weight, see 9.9.9.5. Floor loading criteria to be provided, together with allocated space and height requirements. Access availability to be identified.

Chargers

The enquiry procedure for chargers is in two parts as follows:

(a) Charger size, see section 9.9.9.3 (a) to (d).

(b) Charger dimensions, including weight.

9.9.9.5   Battery accommodation

The battery accommodation required is determined by the size and type of the installation. The types of installation are as follows.

Central battery installations

The battery power is centralised in one room. Conventional batteries can be installed in tiers which in themselves can be used individually or in combination. The tier arrangement for conventional batteries allows for the cells/units to be vertically offset when used in combination to allow access for maintenance. Topping-up is not necessary on valve regulated types and therefore the layouts can be more compact and depending upon floor loadings can rise up to eight shelves (tiers). The stands are generally of modular construction, expandable to accommodate various sizes of battery.

A central battery room should be designed for that specific purpose in respect of ventilation, corrosion environments, and safety.

Office-compatible installations

In this type of installation the batteries are installed within the office environment. This type of battery layout is quite simple, employing fairly high energy density storage and normally using valve regulated batteries. Such cabinets are very flexible in terms of depth and each shelf can accommodate one or two blocks or a single large, front-connected cell.

The office compatible installation does not require special emphasis on ventilation or corrosion environments. However, safety issues still apply.

Self contained systems

Typical of this type of system is single point emergency lighting and alarm equipment, in which the battery is contained in the same enclosure as the charger and the end load. The batteries are usually low voltage and low capacity.

System-free installations

This type of installation covers those battery installations that may be found in buildings, but whose function is more in the nature of a working battery rather than a standby battery. Typical of these installations are batteries for starting stationary engines, and batteries for mechanical handling equipment. Both types of installation require facilities for battery charging within the building.

9.10   UPS installation

9.10.1   Space requirements

The location of any UPS system is important and it should preferably be in a position where it can be accessed quickly and easily in the event of any malfunction.

However, it should primarily be secure, offering resilient support to the critical load. At the concept stage the following points should be considered:

—   That the equipment can be installed at any stage of the building works and that any future expansion requirements can be met. Large scale dismantling will be expensive.

—   Consideration should be given to the ability to install heat extraction systems and provide cooling and combustion air for diesel driven prime movers, if used.

—   Structural loadings for modules, batteries and supporting plant and switchgear should be considered and the effect of noise and vibration investigated for diesel driven systems.

—   Avoid locations in confined low level spaces where flood water could rapidly build up. Spaces with charged piped services should also be avoided.

—   For resilience under battery operation the space should be as high as possible to increase room volume such that the ambient temperature will not rise too fast if cooling is not provided.

—   Input and output switchgear should preferably be local to the UPS equipment and easily identifiable. Consideration should be given to separation to allow access for manual switching in the event of the operation of any automatic fire suppression.

—   Noise emission should be considered from the initial stages. This is applicable to all systems from small dedicated units up to large multiple diesel rotary systems. Clearly this is important where units are proposed for office or computer room environments. The effect on maintenance staff of high noise levels from large diesel prime movers in plant rooms should not be overlooked.

—   In order that HVAC plant failure or maintenance has minimal effect on the UPS operation, consideration should be given to providing independent access to plant located around the periphery of the UPS space; standby units should also be provided if possible.

—   Separation of UPS modules and their batteries should be considered for large systems in order to prevent excessive loss of extinguishant, in the event of accidental discharge. This also provides the opportunity to maintain lower battery operating temperatures, thereby extending battery life.

9.11   UPS testing and commissioning

9.11.1   Specification

The testing and commissioning activities for UPS systems can be an intensive activity employing the use of specialist engineers and equipment. In small dedicated systems the activity is likely to be repetitive with similar units being procured for similar office equipment and as such sample testing may be acceptable. On large central units used to support PDUs or central building systems the tests will be unique and will probably need to be carried out both in the factory and on site when the installation is completed. Factory tests on systems above 1 MV·A are generally only possible with linear loads since it is impractical to recreate a general, or particular, non-linear load.

For tender comparison purposes and to prove that systems are as specified it is essential that the testing activities required are clearly stated at the time of enquiry.

9.11.2   Initial inspection and testing

Initial inspection and testing of UPS equipment should embrace and prove various manufacturers' performance statements but should cover the following:

— suitability of the equipment to be installed within the space provided

— ability of the equipment to provide uninterrupted conditioned power to the dedicated load under variable input conditions of voltage and frequency

— harmonic loads and their effect on the output voltage waveform

— harmonic current content in the input current at various load conditions

— efficiency checks at various load levels including unbalanced loads

— autonomy tests at specified loads and part loads if possible (this will only be possible as an on-site test for larger equipment)

— continuity of supply under various fault conditions within the equipment

— effects of continuous overloads and transient overloads on the system in mains

— 'healthy' and 'failed' conditions

— acoustic tests under various load conditions.

Initial inspection and testing of small dedicated units may take the form of equipment loans prior to purchase, to enable analysis in the actual 'field' conditions. In this case, unless controlled generator power is available, it will be difficult to test the quality of output under difficult input conditions.

On systems of 5 kV·A and upwards the equipment is not easily transportable due to the weight and size of supporting batteries and, once the selection process has occurred, the first tests will be just prior to actual delivery.

The level of initial testing, inspection and assessment of UPS equipment in the 1 kV·A to 20 kV·A band becomes a question of commercial viability and risk. Without major upheaval it is difficult to test in 'field' conditions but factory testing may be costly. The designer and client can only assess the risk.

If factory tests are decided upon prior to delivery of the equipment specified, it should be recognised that, for systems using batteries as the stored energy medium, they are not always sourced from the UPS factory. Often they are delivered directly to site and erected there. This means that factory tests will probably be carried out with a central DC supply.

The majority of manufacturers' test bays are equipped with high speed, computerised data gathering units that provide processed information in a standard form. These can generally print waveforms of voltage and current characteristics.

The more tests specified the higher will be the cost of testing, which is not significant on larger systems but can be high for systems less than 50 kV·A. For specification purposes, the manufacturer should be asked to provide some or all of the following equipment and instruments to be used for testing purposes:

— current, voltage and power factor monitoring equipment for both input and output

— current and voltage monitoring equipment for DC limb

— high sensitivity digital frequency meters

— line spectra visual display instrument with print-out, set to measure percentage current harmonics.

— oscilloscope with variable timebase and hard copy read-out to record current and voltage oscillograms on all phases

— multi-stage three-phase resistive load bank with independent adjustment of each phase

— a separate UPS with 6-pulse rectifier to provide a distorted test load.

For each test condition the following readings should be taken as a minimum:

— input line voltages

— input phase voltages

— fundamental (50 Hz) input currents per phase

— input phase power (kW) or phase power factors

— input harmonic currents for orders 2 to 23

— output line voltages

— output phase voltages

— fundamental (50 Hz) output currents per phase

— output phase power (kW) or phase power factors

— output harmonic voltages for 2nd, 3rd and 5th orders.

Typical tests should be carried out for the following load conditions:

— no load tests

— 50% full load balanced at unity power factor

— unbalanced load with 50%, 100%, 100% full load phase values

— unbalanced load with 25%, 50%, 50% full load phase values (for multi-module systems)

— mains failure with above load(s) (oscillogram)

— removal of above load(s) and return (oscillogram)

— full load tests

— mains failure at full load (oscillogram)

— removal and return of full load (oscillogram)

— operation of automatic bypass (oscillogram)

— 150% full load and time for reversion to bypass

— three phase short circuit test to show fuse clearance.

The above tests are suggestions only and may require adjustment to suit the exact equipment in question. Other adjustments may involve the acceptance of computerised voltage, current and frequency plots in lieu of full oscillograms.

9.11.3   Bringing into service

Once equipment has been delivered to site and installed in its operational mode, including all fire systems or batteries as applicable, a further set of tests should be carried out. To maintain clear responsibility it is recommended that the manufacturers or their agent be employed to install all equipment from system input terminals up to the main outgoing switchgear connection.

Initially the manufacturer should carry out no-load tests to check phase rotation, emergency and safety circuits, alarms, instruments and controls. Only upon completion should the system be accepted for witness testing, as follows:

— full and partial load tests on mains

— full load test with mains interrupted

— for battery stored energy system, load take-up by standby generators

— for battery stored energy systems, autonomy tests for each individual module in parallel configurations, then as a complete system

— functional switching and operational checks

— for systems using diesel back-up, fuel consumption and engine function tests should be carried out at various load levels.

To enable satisfactory testing without affecting the critical load (or in its absence), the need for on-site test load banks should be assessed for each project.

Prior to delivery to site certain vital functions should be achieved. The room should obviously be secure and finished and subject to the needs of the equipment the room finishes sealed to prevent dust migrating to the area.

Subject to the level of financial risk and statutory requirements the room space should be provided with all necessary fire protection, which may need to be operational from the time of permanent operation.

Any necessary air conditioning or ventilation should be checked, commissioned and available from the time that the UPS system goes into prolonged test mode or permanent operation.

Prior to final connection of the UPS system to the critical load the system should go through a prolonged load test of up to 100 hours continuous unsupervised operation, although this is normally achieved during the computer system tests as the whole installation is worked up.

9.12   Operation and maintenance

9.12.1   Operating procedures and instructions

Operating and maintenance instructions should be a composite part of any enquiry sent to manufacturers in order that they make sufficient allowance for their provision. Primarily the documentation should incorporate a list of each fundamental piece of equipment supplied along with serial numbers, ratings and operating voltages and frequency.

Safety precautions should be included as a section within the documentation and should give checking procedures for isolation switching, bypass switching and for initial energisation of the component and load. Further safety warnings should be included regarding hazardous materials and the presence of charged capacitors, if they exist, within the specified equipment.

Where applicable, and particularly in the case of large parallel redundant systems, details of all connections, interconnections and assembly procedures should be provided along with terminal arrangements for monitoring, alarm and diagnostic connections.

Where programmable controllers are used to monitor operational functions, failures and diagnostic information its functions, features and method of use should be indicated.

A general description section should be incorporated including illustrations of major component parts, technical data and dimensions, together with drawings. For completeness and the assistance of attending maintenance engineers the operating instructions should incorporate detail drawings of each major component part including batteries, stands and diesel engines where applicable. Full service parts lists should be provided for each component part including any fault finding procedures.

Finally the operating and maintenance manual should incorporate regular maintenance schedules and procedures, together with test results carried out on completion of manufacture, witnessed in the factory, or subsequent site tests after installation.

9.12.2   Metering, indications and diagnostics

The degree of monitoring and diagnostic information required by the user will be a function of the critical load served and the emergency support facilities that are available on site. In general the information will take the form of liquid crystal integral displays with a select button typically giving the following information:

— input voltage

— output voltage

— input frequency

— output frequency

— input current

— output current

— output load kW and power factor

— normal operation

— load on bypass

— battery voltage

— time of load on battery autonomy

— end of autonomy

— supply voltage and frequency faults

— over-temperature

— event recorder.

In addition, data interface and volt-free alarm contact facilities should be available for remote indication of normal operating conditions and alarms.

For monitoring purposes consideration should be given to the provision of a complete remote monitor unit displaying all of the functions listed including phase values where necessary and allowing interrogation from the remote unit.

A very useful addition to any UPS output switchboard is the provision of built-in test points with integral current transducers, with fully protected outputs, for the periodic attachment of measuring devices including harmonic analysers to keep track of the load profile development and any characteristic changes.

For larger installations extending to parallel redundant arrangements, the operational and status indications should be available both on a per module basis, incorporated on the module enclosure, and on a group basis available on the output/bypass cubicle. In addition mimic diagrams should be available showing the route by which energy is fed to the critical load.

In large central UPS installations it is likely that the equipment is very remote from the critical load centres and in this situation the design should give consideration to remote status panel(s) located in agreed positions and arranged to give separate indication for each module of the following:

— normal operation

— module on battery

— module alarm

— module overload

— low battery volts

— battery fully discharged

— input and output out of synchronism

— system on bypass.

The panel should incorporate visual and audible alarm facilities with 'mute', 'acknowledge test' and 'reset' functions and should, where applicable, provide further indication of emergency or normal power status or diesel generator(s) which have failed to start. This latter status

indication is essential in order that the benefit of battery autonomy can be optimised by the load operator.

Where central BMS systems are installed the designer should give consideration to the provision of an interface unit to give the following possible indications of the system:

— incoming mains voltage

— incoming mains current

— resultant input kV·A

— output voltage

— output frequency

— common fault

— battery operation

— load on bypass

— time elapsed on battery operation

— theoretical time before total discharge

— monitor status of auto bypass and manual bypass.

In large central installations it is likely that, to comply with client requirements or those of the insurers, the designer will need to incorporate automatic gaseous extinguishing into the UPS/battery rooms. The medium used will be subject to current availability/acceptability but with low gas/air concentrations the cause of any fire should be removed before extinguishing action. To achieve this a fire shut-down signal can be introduced to de-energise the input and output to the UPS. Test facilities should be available to allow a fire test to be carried out without de-energising by the inclusion of a 'fire test inhibit' switch.

To assist with operational switching in the event of fire, consideration should be given in the early project planning stage to the location of switchgear outside, but adjacent to, the UPS equipment room.

Where diesel driven UPS equipment is proposed the various local and remote monitoring facilities should include indication of the following:

— high diesel water temperature

— low diesel oil pressure

— diesel failed to start.

This last signal is applicable for battery backed diesel UPS only, since several attempts can be made whilst on battery. In the case of kinetic energy storage systems, upon mains failure, it means that the critical load has been switched off.

9.12.3 Remote control, diagnostics and service

Many UPS systems are now produced with the ability to communicate over a telephone line for remote service and fault diagnostics. Modules can be programmed so as to call a pre-arranged number and report a fault. If this is a requirement then the designer must allow for the installation of a dedicated (direct) line to the UPS module. In addition, depending upon the manufacturer of the UPS, it may be required to install a modem and 13 A socket local to the module. Additional 13 A single phase power outlets may be required for external printers, service PCs and instruments.

References

1. Council Directive of 3 May 1989 on the approximation of the laws of the member states relating to electromagnetic compatibility EC Directive 89/336/EEC *Official J. of the European Communities* 23.5.89 L139/19 (as amended by Council Directives 92/31/EEC (1992), 93/68/EEC (1993) and 93/97/EEC (1993)) (Brussels: Commission of the European Communities) (1989–93)

2. *Planning levels for harmonic voltage distortion and the connection of non-linear equipment to transmission systems and distribution networks in the United Kingdom* Engineering Recommendation G5/4 (London: Electricity Association) (2001)

3. BS 6290: *Lead-acid stationary cells and batteries*; Part 2: 1999: *Specification for the high-performance Planté positive type*; Part 3: 1999: *Specification for the flat positive plate type*; Part 4: 1997: *Specification for classifying valve regulated types* (London: British Standards Institution) (1997, 1999)

4. Falk S U and Salkind A J *Alkaline storage batteries* (New York: John Wiley) (1969)

5. BS 6133: 1995: *Code of practice for safe operation of lead-acid stationary batteries* (London: British Standards Institution) (1995)

6. BS 6132: 1983: *Code of practice for safe operation of alkaline secondary cells and batteries* (London: British Standards Institution) (1983)

7. BS EN 60896: *Stationary lead-acid batteries. General requirements and methods of test*; Part 1: 1992: *Vented types*; Part 2: 1996: *Valve regulated types*; Part 11: 2003: *Vented types. General requirements and methods of test* (London: British Standards Institution) (1992–2003)

8. BS EN 60254: *Lead-acid traction batteries*; Part 1: 1997: *General requirements and methods of test*; Part 2: 1997: *Dimensions of cells and terminals and marking of polarity on cells* (London: British Standards Institution) (1997)

9. IEC 60254: *Lead-acid traction batteries*; Part 1: 1997: *General requirements and methods of test*; Part 2: 2000: *Dimensions of cells and terminals and marking of polarity on cells* (Geneva: International Electrotechnical Commission) (1997, 2000)

10. BS EN 60095: *Lead-acid starter batteries*; Part 1: 1993: *General requirements and methods of test*; Part 4: 1993: *Dimensions of batteries for heavy commercial vehicles* (London: British Standards Institution) (1993)

11. BS EN 61056: *Portable lead-acid cells and batteries (valve-regulated types)*; Part 1: 2003: *General requirements, functional characteristics. Methods of test* (London: British Standards Institution) (2003)

12. IEC 61056: *General purpose lead-acid batteries (valve-regulated types)*; Part 2: 2002: *Dimensions, terminals and marking* (Geneva: International Electrotechnical Commission) (2002)

13. IEC 60623: 2001: *Secondary cells and batteries containing alkaline or other non-acid electrolytes. Vented nickel–cadmium prismatic rechargeable single cells* (Geneva, Switz: International Electrotechnical Commission) (2001)

14. IEC 60622: 2002: *Secondary cells and batteries containing alkaline or other non-acid electrolytes. Sealed nickel–cadmium prismatic rechargeable single cells* (Geneva: International Electrotechnical Commission) (2002)

15. BS EN 61150: 1994: *Alkaline secondary cells and batteries. Sealed nickel–cadmium rechargeable monobloc batteries in button cell design* (London: British Standards Institution) (1994)

16. *IEEE Recommended practice for sizing lead-acid batteries for stationary applications* ANSI/IEEE Standard 485 (New York, NY: Institute of Electrical and Electronics Engineers) (1997)

10   Earthing

10.1   Introduction

Earthing is the process of connecting to the general mass of earth by means of electrically conducting materials those conductive parts of an installation which are not normally subject to a voltage or electrical charge. In some instances earthing may be applied also to certain conductive parts which will normally carry a small and controlled voltage or charge.

Bonding is the related process of interconnecting such conductive parts to ensure that even under fault conditions the voltage differences between adjacent conductive parts are restricted to safe levels. It should be noted that bonding does not ensure that adjacent surfaces remain at the same potential under all conditions.

Earthing and bonding have three main objectives:

— *Safety*: which is achieved primarily by limitation of the voltage which may appear on conductive surfaces under fault conditions and by limiting the period of time for the condition to persist.

— *Protection of buildings, plant and equipment*: which is achieved partly by voltage limitation but mainly by ensuring that fault currents are sufficiently large to produce effective reaction times in protective devices.

— *Correct and precise operation of equipment*: which is achieved by providing a constant and noise-free electrical reference plane.

10.2   Statutory requirements, standards and definitions

10.2.1   Statutory documents

The Electricity at Work Regulations 1989[1], of which Regulations 8 and 9 deal specifically with earthing and the integrity of earthed conductors.

10.2.2   British Standards

Relevant British Standards are as follows:

— BS 7430: 1998: *Code of practice for earthing*[2]

— BS 6651: 1999: *Code of practice for protection of structures against lightning*[3]

— BS 7671: 2001: *Requirements for electrical installations. IEE Wiring Regulations. Sixteenth Edition*[4]

— BS 7361: Part 1: 1991: *Cathodic protection. Code of practice for land and marine applications*[5]

10.2.3   Related standards

Related codes and standards are as follows:

— *Distribution Code of the Public Electricity Suppliers of England and Wales*[6]

— Engineering Recommendation G59/1: *Engineering recommendations for the connection of embedded generating plant to the RECs' distribution systems*[7]

— Engineering Technical Report ETR113: *Notes of guidance for the protection of private generating sets up to 5 MW for operation in parallel with PES distribution networks*[8].

10.2.4   Definitions

Definitions used in this section are in accordance with BS 7430[2], BS 6651[3] and BS 7671[4], as appropriate. Illustrations of some of the more commonly used terms are given in Figures 10.1 to 10.4[2,4,9].

10.3   Design principles

10.3.1   General principles

By convention the general mass of earth is taken to be at zero voltage. Connecting the neutral point of an electricity supply system to earth therefore creates a reference plane against which all system voltages are measured and limits the possible rise in potential of any conductive surface, see Figure 10.5.

The mass of earth, however, is not a sink. Current flow into it from an installation will flow out at the supply point neutral earth connection(s), to complete the current circuit. Because it is a reasonably good conductor the earth performs this function quite well. Enhanced performance is obtained by the provision of suitable protective conductors between conductive surfaces and the source neutral, as much lower and more controllable circuit impedances can be achieved.

10.3.2   Scope

This Guide is concerned mainly with low voltage systems at normal mains frequency. High voltage considerations are made where these impact upon the low voltage system.

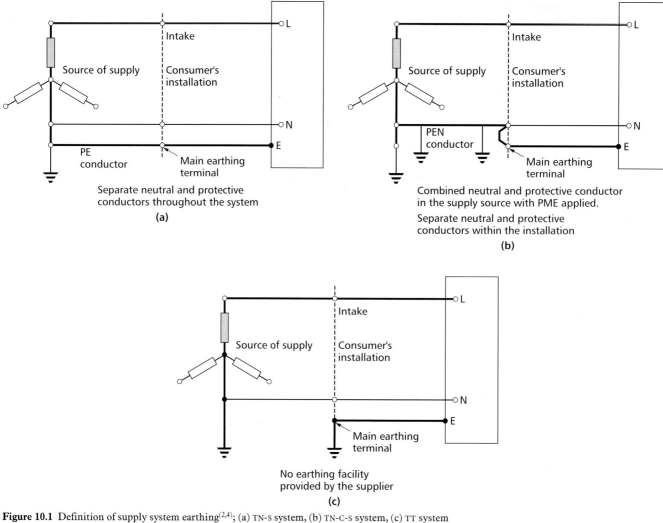

Figure 10.1 Definition of supply system earthing[2,4]; (a) TN-S system, (b) TN-C-S system, (c) TT system

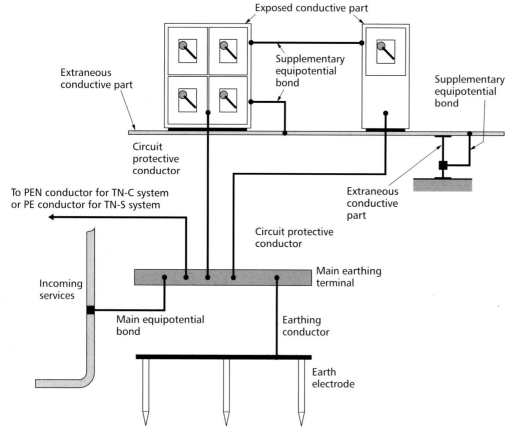

Figure 10.2 Definition of earthing terminology[4]

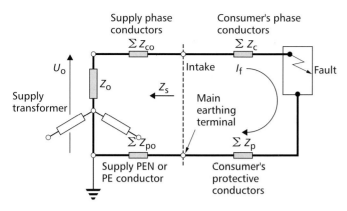

Z_s = Supply earth loop impedance = $Z_o + \Sigma Z_{co} + \Sigma Z_{po}$

Z_{el} = Earth loop impedance = $Z_o + \Sigma Z_{co} + \Sigma Z_c + \Sigma Z_p + \Sigma Z_{po}$

(a)

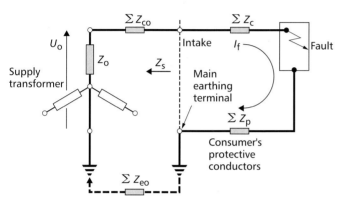

Z_{el} = Earth loop impedance = $Z_o + \Sigma Z_{co} + \Sigma Z_c + \Sigma Z_p + \Sigma Z_{po}$

ΣZ_{eo} includes both electrodes and resistance of the earth mass

(b)

Figure 10.3 Definition of earth loop impedance[9]; (a) TN system, (b) TT system

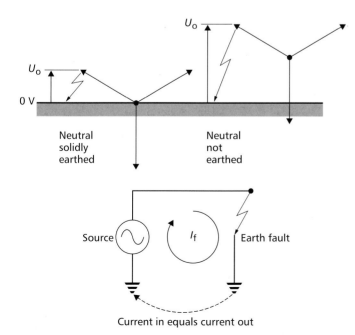

Figure 10.5 Effect of earthing the supply source[4]

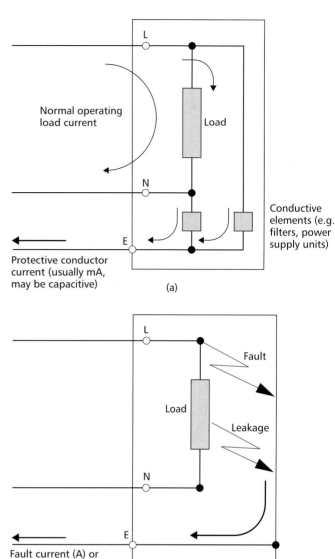

Figure 10.4 Definition of protective conductor and leakage currents[4]; (a) protective conductor current (for special requirements see BS 7671, section 607), (b) fault current or leakage current

10.3.3 Supply systems

The supply systems defined in BS 7430[2] and BS 7671[4] and illustrated in Figure 10.1 are coded in terms of their neutral–earth arrangements.

An IT system has an isolated or impedance-earthed neutral point with no distributed protective conductor. This form of supply is not permitted on public supply networks and is therefore not considered further in this guide.

A TT system has an earthed neutral point but the protective conductor is not distributed. The installation must therefore be earthed by an electrode system at the point of intake.

TN systems also have the neutral point solidly earthed but the neutral conductor is distributed either as a combined neutral/protective conductor in which case it is a TN-C system or, if it has separate neutral and protective conductors, a TN-S system. Where a TN-C supply is provided it is commonly carried through the installation as TN-S, in which case the whole system is described as TN-C-S. This is the common form of protective multiple earthing (PME). The combined protective and neutral conductor is designated PME but is also referred to as the combined neutral and earth (CNE) conductor or protective earthed neutral (PEN) conductor.

The PEN conductor is earthed at several points along its length. An electrode is normally provided close to the intake point.

Since the PEN may be common to several consumers it is evident that a voltage impressed on it from one consumer could be imported into the premises of another. For this reason it is a requirement that all metal pipework services entering the premises are bonded at the earliest point onto the consumer's main earthing terminal.

10.3.4   High voltage supplies

A public supply at high voltage would normally be a three-phase three-wire system. Cable armouring and screens would be earthed at least at the sending end and also probably at the intake point. It is a requirement of the *Distribution Code*[6] that the public electricity supplier (PES) or regional electricity company (REC) makes available details of the earthing arrangement and values of earth loop impedance at the point of connection.

It is not a requirement of the code that earthing facilities for a consumer's installation should be provided by the PES. A high voltage earth electrode system must therefore be provided by the consumer with an earthing conductor to a main earth terminal. Exposed conductive parts of the consumer's high voltage equipment must be connected to the main earth terminal with suitably rated circuit protective conductors (CPCs). These connections include:

— cable armouring

— cable screens

— switchgear metal framework and panelwork

— transformer casings and extraneous metalwork

— transformer screens

— transformer metal enclosures

— transformer iron core laminations.

For oil filled transformers the screens, laminations and frame are usually bonded together and a substantial earthing stud is provided. For cast resin transformers screens are provided only to special order and although these are normally provided on the outside of the low voltage winding they must nevertheless by bonded to the high voltage earth system.

Where a combined earth electrode resistance not exceeding 1 Ω can be achieved a common HV/LV system is permitted. More commonly the resistance will be between 5 and 10 Ω. In such cases separate electrode systems must be provided (see section 10.3.5) but, as it is almost impossible to provide electrical separation between the incoming supply, the consumer's HV network and the LV network, it is essential that all three are interlinked. The bonding links should be equal in size to the largest HV circuit protective conductor (CPC).

10.3.5   Electrodes

10.3.5.1   Soil

The electrical resistivity of the soil has a major influence on the effectiveness of an electrode system. Many of the factors affecting soil resistivity vary locally and even seasonally. Such factors as moisture content, soil structure, chemical compositions and the nature and persistence of any salts will have a major impact. Temperature by contrast has a marked effect only at or around freezing point. For this reason and because of the variable height of any water table it is often advisable to ignore the first metre depth of a driven electrode.

Consideration of the type of soil and any stratification will dictate the most suitable form of electrode. For example a shallow underlying rock stratum will probably lead to the use of horizontal buried electrodes, whereas a high resistivity gravel layer with underlying clay would lend itself to deep driven rods.

10.3.5.2   Electrodes

Extensive detail relating to electrodes is contained in section 3 of BS 7430[2].

The most commonly used electrode system is in the form of copper or copper clad steel rods interconnected with copper tape. In general, and for uniform soil structures, little extra benefit is obtained by driving very deep rods. At effective depths of more than about 3 metres the decrease in resistance is relatively small, see Figure 10.6[2]. It is better to use a large number of short rods rather than fewer long ones, unless the low resistivity soil is at the greater depth.

Rods should not be driven into disturbed ground. One of the most economical methods for installing electrodes is to use the compacted base of cable trenches to gain maximum effect from electrodes and connecting tapes.

10.3.5.3   Design

The theoretical design of electrode systems is extremely complex although it can be attempted by reference to the tables and formulae contained in BS 7430[2]. A more pragmatic approach is to design the system based on the data contained in Figure 10.6 to produce a first order approximation. The installation should then proceed from this base by trial and measurement until a satisfactory design value is achieved.

A practical earth electrode system is likely to have a resistance to earth closer to 10 Ω than 1 Ω. Striving for the latter value is often both uneconomical and unnecessary.

Since the earth loop impedance is determined mainly by the contribution from phase and protective conductor impedances, preoccupation with achieving very low electrode resistance represents misplaced effort. As a general rule 10 Ω should be regarded as a guide level with 20 Ω as a 'not to be exceeded' value.

Separation of electrode systems, e.g. high and low voltage systems, is required unless a 1 Ω level can be assured.

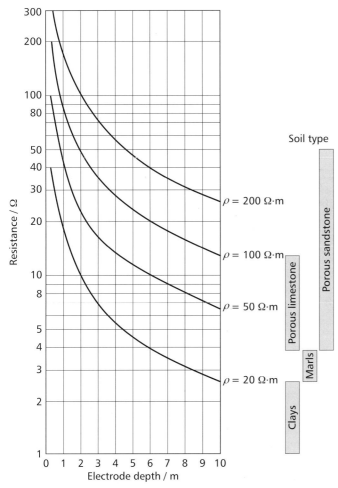

Figure 10.6 Theoretical curves of electrode resistance[4]

Separation means that each system should be outside the resistance area of another which in turn means that the low voltage electrode must be sufficiently far away so as not to be affected by the potential gradient surrounding the high voltage electrode.

10.3.5.4   Materials and corrosion

Cathodic protection is normally applied to wholly or partly buried ferrous structures in order to counteract electrolytic corrosion.

It is unfortunate that low resistivity soils tend to have the most aggressive and corrosive effect on electrode materials. For this reason it is recommended that in low resistivity soils, i.e. in the order of 40 Ω·m, only copper or copper clad steel elements are used. Solid copper is particularly suitable for high fault current situations.

Connections between dissimilar metals can have possible galvanic effects. In general connection of galvanised steel to steel in concrete should be avoided, as should galvanised steel to copper. In contrast, the connection of copper to steel in concrete is normally safe.

Aluminium or copper clad aluminium should not be used in contact with soil or in damp conditions and should not be used to make the final connection to an earth electrode.

10.3.5.5   Miscellaneous electrodes

Since the introduction of polymeric oversheaths, metallic cable sheaths no longer provide the fortuitous earth connection they did in former times. Cable armouring may still be employed as a protective conductor, either alone or associated with a separate circuit protective conductor (CPC).

Structural steelwork, sheet steel piling and steel reinforcement of concrete piling can all provide very low resistances to earth. Corrosion of steelwork within concrete is unlikely to be caused by alternating currents. However, if significant levels of leakage current may reasonably be anticipated, particularly where a direct current component may be present, then the steelwork should at best be regarded as an auxiliary electrode and bonded to a main electrode system. Care must be taken not to introduce corrosion-producing currents by connecting incompatible materials to the building steelwork, see 10.3.5.4 above. It should be remembered that as the concrete and ground dry out so the resistance will increase.

10.3.5.6   Water and other service pipes

Other metallic service pipes should not be used as earth electrodes. They must however be bonded to the main earth terminal at the point of entry into the building and may act as an auxiliary path to earth.

10.3.5.7   Cathodically protected structures

Cathodic protection works by maintaining a building's ferrous structure at a slightly negative potential with respect to earth. Connection of an earth electrode to such a structure may increase the current drain from the protective source and may therefore be unacceptable. A connection may nevertheless be required for protective and functional reasons and in such cases careful consideration must be given to the form of earth electrode. BS 7361: Part 1[5] provides detailed guidance.

Copper should not be connected directly to the ferrous structure. Zinc may be used in some instances or austenitic stainless steel plus a moderate earth resistance to produce an acceptable increase in current drain.

10.3.6   Earthing conductors

It is recommended that the earthing conductor should be rated to carry the full calculated earth fault current regardless of any parallel paths and auxiliary electrodes that may exist.

The duration of the current should be taken as one second minimum (preferably three seconds) and not as the reaction time of any protective devices.

It will normally be found that solid copper strip provides the most suitable conductor although on occasion steel strip may be required. Aluminium strip is not recommended for installation direct in the ground or directly onto the electrode.

Figure 10.7 shows graphically the current densities sustainable for one second from an initial temperature of 30 °C for copper and steel to give various final temperatures. Temperatures above 500 °C are not recommended. Above 200 °C, the conductor must be visible throughout

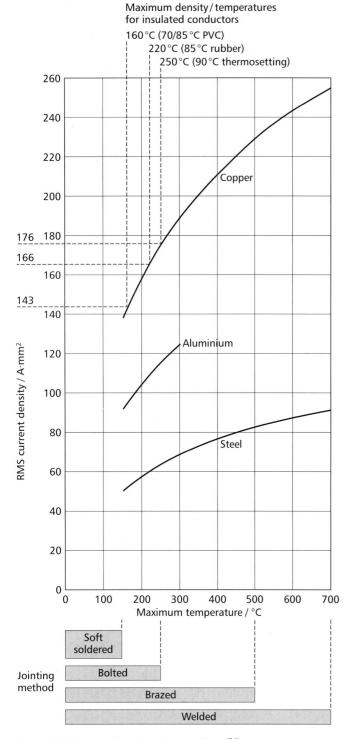

Figure 10.7 Protective conductor current limits[2,4]

Table 10.1 Maximum earth fault currents for various copper sections

Section	Maximum fault current (/ kA) for stated time	
	1 second	3 seconds
20 mm × 3 mm	9.5	5.5
25 mm × 3 mm	11.9	6.9
25 mm × 4 mm	15.9	9.2
25 mm × 6 mm	23.9	13.8
31 mm × 3 mm	14.8	8.5
31 mm × 6 mm	29.6	17.1
38 mm × 3 mm	18.1	10.5
38 mm × 5 mm	30.2	17.4
38 mm × 6 mm	36.3	20.9
50 mm × 3 mm	23.9	13.8
50 mm × 4 mm	31.8	18.4
50 mm × 6 mm	47.7	27.5

—   the earthing conductor

—   outgoing circuit protective conductors

—   main equipotential bonding conductors

—   bonding conductors to other earth electrode systems, e.g. high voltage system and lightning protection system

—   bonding conductors to the building structure.

—   functional (clean earth) protective conductors.

On larger installations the terminal will usually take the form of a section of flat copper bar, suitably sized to carry full fault current and to be of sufficient length and section to enable connection of all of the required protective conductors without impairment of its current carrying capacity. The terminal or flat conductor should be separated from the building structure by robust insulating supports, both to ensure that any temperature rise will not damage the structure and to allow termination of the protective conductors. Ideally all terminated protective conductors should be labelled.

On small installations the terminal may simply comprise a proprietary solid copper or brass connector block, suitably drilled and tapped.

10.3.8   Circuit protective conductors

The function of circuit protective conductors (CPCs) is to carry earth fault currents from the seat of the fault back to the source. This must be achieved in complete safety. Therefore the CPC must be of sufficient size to carry the fault current without damage to itself and must be of sufficiently low impedance in relation to the phase conductor feeding the fault so as to limit the rise in voltage at the conducting surfaces in the vicinity of the fault.

The CPC may consist of cable armouring and/or screens, metal containment systems, separate conductors or a separate core within a multicore cable.

It should be noted that there are only two situations where a specifically separate CPC is prescribed by the regulations:

its length and be suitably supported and separated from organic materials. For temperatures above 150 °C certain building materials may be at risk.

Table 10.1 shows the maximum earth fault currents for one second and three seconds duration in various copper sections for a final temperature of 200 °C (from an initial temperature of 30 °C).

10.3.7   Main earthing terminal

The main earthing terminal is the point at which are connected:

—   the incoming protective conductor

— through flexible or pliable metal conduit

— in a final circuit intended to supply equipment producing a protective conductor current in excess of 10 mA.

Conduit and trunking systems can provide adequate protection at the final circuit end of an installation although much reliance has to be placed on the mechanical integrity of the connections.

For those situations where cables of large cross section are employed, the ratio of conductances between a phase conductor and the cable armour becomes less favourable as the cable section increases. Calculations are required to determine the earth fault loop impedance and hence the fault current to ascertain the suitability of the armour as a CPC. Cable armour is unlikely to produce a suitable limit to the touch voltage at the seat of a fault since, at best, its effective conductance will be less than half that of the phase conductor. The touch voltage therefore would rise to something in excess of 160 V, see section 10.3.9 below.

In general therefore it will be necessary or desirable to provide separate CPCs at the heavier end of an installation to supplement the cable armours.

Separate CPCs should wherever possible be installed adjacent to their associated cable or phase conductor. This will reduce the loop reactance of the earth loop. Table 10.2 (see section 10.4) gives impedance values for various cables sizes. Note that for the smaller cables reactance can be ignored and that, for armouring, only the resistance has been considered. Calculation methods are considered in section 10.4.1.

By installing the separate CPC as described above, complex calculations of earth loop reactance can be avoided and the loop reactance can be taken as approximating to the line-neutral loop reactance for single core unarmoured cables, taking the average of the phase conductor and CPC sizes.

BS 7671[4] provides the following formula for the calculation of the cross sectional area of a CPC:

$$S = \frac{\sqrt{I^2 t}}{k} \qquad (10.1)$$

where S is the cross sectional area (mm^2), I is the prospective earth fault current (A), t is the duration of the fault (s) and k is a factor equivalent to the permitted current density for a given temperature rise for a particular conductor material.

For a duration of one second the equation becomes:

$$S = I / k \qquad (10.2)$$

For a duration of three seconds the equation becomes:

$$S = \sqrt{3} \times (I / k) \qquad (10.3)$$

The value of k for various materials is given in BS 7671[4] or can be taken from the current density (A/mm^2) axis of Figure 10.7.

10.3.9   Touch voltage

IEC 60364: Part 1[10] includes a graph of voltage against time which represents the maximum time that a voltage should persist on a conducting surface under fault conditions to minimise the risk of fatal shock. This graph is reproduced here as Figure 10.8.

An estimate of touch voltage at any point in an installation can be derived from the ratio of earth return path impedance to total earth loop impedance (see section 10.4), i.e:

$$U_t = (U_o Z_p) / Z_{el} \qquad (10.4)$$

where U_t is the touch voltage (V), U_o is the supply phase to neutral voltage (V), Z_p is the return path impedance (Ω) and Z_{el} is the earth fault loop impedance (Ω).

In situations where the earth loop impedance of the supply is reasonably low, a good estimate of the touch voltage will be obtained from a consideration of the consumer's phase conductors and CPCs at each conductive surface. It can be shown from the formulae included in section 10.4 that, for example, where the conductance of the CPC is half that of the phase conductor the touch voltage will approximate to two thirds of the nominal supply phase voltage, i.e. 160 V for a 240 V system.

From Figure 10.8[10] it can be seen that a 160 V touch voltage corresponds with a duration of only 0.1 seconds

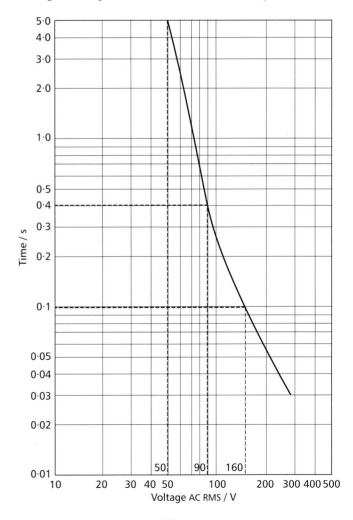

Figure 10.8 Touch voltage curve[10] (reproduced from IEC 60364 by permission of the International Electrotechnical Commission)

and the fault interruption should ideally be designed to operate within this time. However, with properly installed equipotential bonding these touch voltages may be significantly reduced and BS 7671[4] states maximum disconnect times of 5 seconds for most fixed equipment and 0.4 seconds for equipment with which people would normally make contact, e.g. portable tools. It must be stressed that these are maximum times and faster times are recommended.

The recommended maximum disconnect times in BS 7671 of 5 s and 0.4 s relate to touch voltages of 50 V and 90 V respectively. These disconnect times are based on the work of International Electrotechnical Commission (IEC) Technical Committee TC64 (Working Group WG9). This work defined a third parameter, 'prospective touch voltage' (U'_c). Experience and calculations showed that U'_c varied between 0.3 U_o and 0.75 U_o depending on the closeness of the fault to the source and on the ratio of phase and CPC cross-sectional areas. Applying mean values to both factors (i.e. 0.8 for proximity, and equal phase and protective conductors) equates to a disconnect time of 0.4 s with a touch voltage of 92 V.

10.3.10   Equipotential bonding

The purpose of equipotential bonding is to ensure that under fault conditions both within and without an installation the potential difference between simultaneously accessible conductive parts, both exposed and extraneous, is limited to a safe level.

It should be noted that equipotential bonding does not ensure that such surfaces will be at the same voltage level under fault conditions.

Bonding of conductive parts emanating from outside the equipotential zone is particularly important since they could otherwise import unsafe charges from an external fault.

Main equipotential bonding refers to direct bonding between extraneous conductive parts and the main earthing terminal. Incoming pipework services, for example, must be bonded at the earliest point of the consumer's distribution system preferably before any branching has taken place. Sections of incoming metalwork not belonging to the customer's installation and/or isolated from it by insulating gaskets etc., should be insulated.

Supplementary equipotential bonding can be applied as considered necessary between:

—   exposed conductive part to exposed conductive part

—   exposed conductive part to extraneous conductive part

—   extraneous conductive part to extraneous conductive part.

An illustration of the requirements of equipotential bonding is given in Figure 10.9.

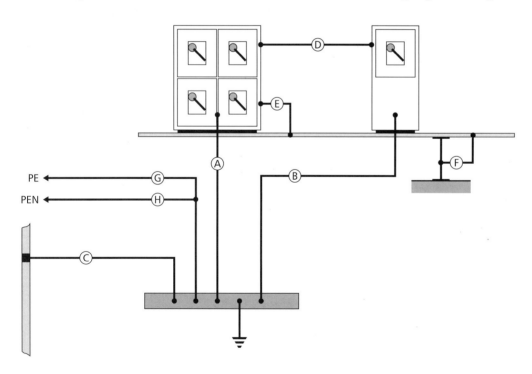

Main equipotential bond for TN-S system: C = G/2 (min. 6, max. 25)

Main equipotential bond C for TN-C system: for H ≤ 35, C = 10
                                               for H > 35 or = 50, C = 16
                                               for H > 50 or = 95; C = 25
                                               for H > 95 or = 150; C = 35
                                               for H > 150; C = 50

Supplementary bond D = B (min. 4 if unprotected)

Supplementary bond E = A/2 (min. 4 if unprotected)

Supplementary bond F = A/2 (min. 4 if unprotected)

Figure 10.9 Equipotential bonding[4]

As mentioned previously the effect of equipotential bonding is to reduce the touch voltage between simultaneously accessible parts since instead of the voltage being governed by the return path impedance back to the energy source it becomes instead effectively the volt drop between the conductive surface at the fault and the main earth terminal, see section 10.4. The impedance of this circuit is clearly within the control of the designer.

10.4 Calculation methods

10.4.1 Fault current

Earth fault current is a function of the system nominal phase voltage and the earth loop impedance at the point in the network under consideration, i.e:

$$I_f = U_o / Z_{el} \tag{10.5}$$

where I_f is the earth fault current (A), U_o is the system nominal phase voltage (V) and Z_{el} is the earth loop impedance (Ω).

The calculations therefore become a matter of assessing the values for Z_{el} throughout the installation.

Conductor impedance data for copper conductors and steel wire armouring are given in Table 10.2. It should be noted that impedance values are in mΩ/m and, if used directly, will produce fault currents in kiloampères (kA).

10.4.1.1 LV systems with intake up to 100 A

See Figure 10.10.

$$Z_{el} = Z_s + \Sigma R_c + \Sigma R_p \tag{10.6}$$

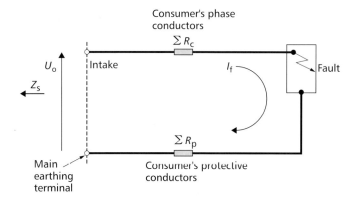

Figure 10.10 Earth fault current calculations[5]; circuits up to 100 A

where Z_{el} is the earth loop impedance at seat of fault (Ω), Z_s is the supply earth loop impedance (Ω), ΣR_c is the combined resistance of the consumer's phase conductors (Ω) and ΣR_p is the combined resistance of the consumer's protective conductors (Ω).

The supply earth loop impedance is as follows:

— TN-C supplies: approx. 0.35 Ω

— TN-S supplies: approx. 0.8 Ω

— or as advised by supply company.

The earth fault current is then obtained from equation 10.5.

Note: the loop reactance $\Sigma (X_c + X_p)$ may be ignored; the calculation of Z_{el} needs only to be arithmetical.

If a separate CPC is used together with cable armouring, the combined resistance is:

$$R_p = \frac{R_{CPC} R_a}{R_{CPC} + R_a} \tag{10.7}$$

Table 10.2 Conductor data

Cross-sectional area / mm²	R_c or R_p / (mΩ·m⁻¹)	$(X_c + X_p)$ / (mΩ·m⁻¹)	R_{a2} / (mΩ·m⁻¹)	R_{a3} / (mΩ·m⁻¹)	R_{a4} / (mΩ·m⁻¹)
1.5	15.6	—	—	—	—
2.5	9.2	—	11.2	10.3	9.6
4	5.8	—	10.0	9.2	8.4
6	3.9	—	9.0	8.1	5.2
10	2.3	—	7.7	5.0	4.5
16	1.4	—	4.2	4.0	3.5
25	0.92	0.31	4.1	2.8	2.5
35	0.66	0.29	2.8	2.5	2.2
50	0.50	0.29	2.5	2.2	2.0
70	0.34	0.28	2.2	2.0	1.3
95	0.25	0.27	1.5	1.4	1.2
120	0.20	0.26	1.4	1.2	0.84
150	0.16	0.26	1.3	0.86	0.75
185	0.13	0.26	0.90	0.78	0.67
240	0.10	0.26	0.80	0.70	0.59
300	0.08	0.25	0.74	0.64	0.54
400	0.07	0.25	—	0.53	—

Note: R_c or R_p = conductor resistance for phase or CPC; $(X_c + X_p)$ = approximate loop reactance for equal size phase conductor and CPC; R_{a2}, R_{a3} etc. = armour resistance for a 2-core, 3-core etc. cable adjusted to allow for magnetic effect

$(X_c + X_a)$ = approximate armour/conductor loop reactance (= 0.3 mΩ·m⁻¹)

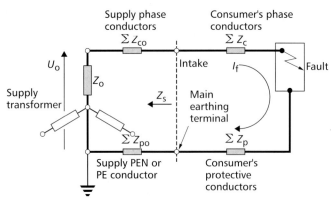

Figure 10.11 Earth fault current calculations[5]; circuits above 100 A

10.4.1.2 LV systems with above 100 A

See Figure 10.11. The earth loop impedance is given by:

$$Z_{el} = Z_s + \Sigma Z_c + \Sigma Z_p \qquad (10.8)$$

where Z_{el} is the earth loop impedance at seat of fault (Ω), Z_s is the supply earth loop impedance (Ω), ΣZ_c is the combined impedance of the consumer's phase conductors (Ω) and ΣZ_p is the combined impedance of the consumer's protective conductors (Ω).

The supply earth loop impedance is given by:

$$Z_s = Z_o + \Sigma Z_{co} + \Sigma Z_{po} \qquad (10.9)$$

i.e:

— TN-C supplies up to 200 A: approx. 0.35 Ω

— TN-C supplies up to 300 A: approx. 0.2 Ω

— TN-C supplies above 300 A: approx. 0.15 Ω

— or as advised by supply company.

For each section of the consumer's network:

$$Z_c + Z_p = (R_c + R_p) + j(X_c + X_p) \qquad (10.10)$$

where R_c is the phase conductor resistance (Ω), R_p is the circuit protective conductor resistance (Ω) and $(X_c + X_p)$ is the approximate loop reactance (Ω).

Assuming Z_s to be predominantly reactive:

$$|Z_{el}| = \sqrt{\left(R_c + R_p\right)^2 + \left[Z_s + \left(X_c + X_p\right)\right]^2} \qquad (10.11)$$

The earth fault current is then obtained from equation 10.5.

10.4.2 Touch voltage

Calculation methods are given below for systems with and without equipotential bonding. For systems where the supply point earth loop impedance is relatively low an appropriate calculation method is given in section 10.4.2.3.

Conductor data should be taken from Table 10.2. Reference should then be made to the touch voltage/time curve in Figure 10.8 to obtain suitable operating times for protective devices.

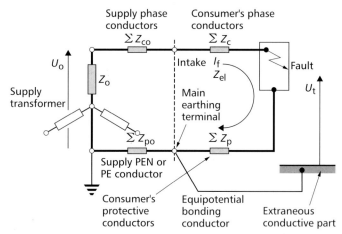

Figure 10.12 Touch voltage calculation[5]

10.4.2.1 Circuits without equipotential bonding

See Figure 10.12. The touch voltage is given by:

$$U_t = I_f \times \Sigma(Z_p + Z_{po}) \qquad (10.12)$$

where U_t is the touch voltage (V) and Z_p is the impedance of the circuit protective conductors (Ω) and Z_{po} is the impedance of the supply PEN or PE conductor (Ω).

Hence, substituting for I_f from equation 10.5:

$$U_t = U_o \frac{\Sigma(Z_p + Z_{po})}{Z_{el}} \qquad (10.13)$$

10.4.2.2 Circuits with equipotential bonding

See Figure 10.12. For circuits with equipotential bonding, equation 10.13 reduces to:

$$U_t = U_o \frac{\Sigma Z_p}{Z_{el}} \qquad (10.14)$$

10.4.2.3 Approximate method

See Figure 10.13. For systems where the supply point earth loop impedance is relatively low, Z_{el} is given by:

$$Z_{el} \approx \Sigma Z_c + \Sigma Z_p \qquad (10.15)$$

Thus, from equation 10.5:

$$I_f \approx \frac{U_o}{\Sigma Z_c + \Sigma Z_p} \qquad (10.16)$$

Touch voltage is given by:

$$U_t \approx U_o - I_f(\Sigma Z_c) \qquad (10.17)$$

Hence:

$$U_t \approx U_o \left(1 - \frac{\Sigma Z_c}{\Sigma Z_c + \Sigma Z_p}\right) \qquad (10.18)$$

i.e:

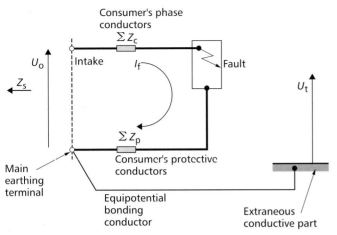

Figure 10.13 Approximate touch voltage calculation

$$U_t \approx U_0 \frac{\Sigma Z_p}{\Sigma Z_c + \Sigma Z_p} \qquad (10.19)$$

If, for example, $\Sigma Z_c = 2 \Sigma Z_p$, then:

$$U_t \approx \tfrac{2}{3} U_0 \qquad (10.20)$$

10.5 Special requirements

10.5.1 Functional earthing

As its name implies, functional earthing is provided to enable equipment to operate properly. Strictly speaking, it does not provide protection either to the equipment or to people using it.

Examples of functional earthing are:

— provision of a zero voltage reference point

— electromagnetic screening

— signalling path for communications equipment.

The functional earth conductor is permitted to be connected to the main earthing terminal either directly or indirectly. It should be installed in accordance with BS 6701: Part 1[11] where used for telecommunication equipment. Where used for other purposes it may again be connected to the main earthing terminal and suitably labelled.

10.5.2 Clean earth

The so-called 'clean earth' often insisted upon by data processor users should not be confused with functional earthing. It is essential that the clean earth conductor does not become isolated from the normal installation earth.

It is common practice to provide five-core armoured cables for data processing equipment. In this case the fifth core is the clean earth and the armour is the CPC. The two conductors should be connected together at every termination point, e.g. the power distribution unit, the local link box and the processing equipment.

The clean earth may be derived from the neutral point of an isolating transformer but this point must be solidly connected to the main building earth.

It must be a rule that the protective features of the earthing system always outweigh any functional requirements.

10.5.3 Data processing installations

Data processing and telecommunication installations are not exempt from the normal safety and protective requirements of earthing systems described above, and BS 7671[4] is quite clear in its statement that each exposed conductive part of a data processing installation must be connected to the main earthing terminal.

Data processing equipment often gives rise to relatively large protective conductor currents. Section 607 of BS 7671 details the earthing requirements for equipment and systems having such currents in excess of 3.5 mA and 10 mA. The requirements include the use of connectors to BS EN 60309-2[12] (formerly BS 4343), the size of the protective conductor and the provision of a second protective conductor terminated separately from the first CPC.

CPCs must be routed with their associated phase conductors, preferably within a multicore laid-up cable or alternatively strapped to the power cable over its whole length. This is to minimise any field which would be set up within the loop area of a separated power conductor and the CPC. In any case it is bad practice to separate the CPC from its power conductor as the increased inductance introduces an unnecessary additional impedance to the earth fault loop.

It has been shown that the supporting structure for a raised floor can serve quite effectively as a high frequency signal reference grid. Bolted-down metal stringers making good contact with steel support pedestals provide the best system. Stringerless systems are also suitable provided that they incorporate good contact springs between the floor plates and the pedestals and conducting coverings. It is recommended that an earthing grid be provided with cable connections to at least one pedestal in every nine. The grid should be connected with main equipotential bonding conductors to the main earthing terminal and supplementary equipotential bonding conductors should be connected between the floor grid and any exposed conductive part.

10.5.4 Lightning protection

BS 6651: *Code of practice for protection of structures against lightning*[3] is a comprehensive document and is used extensively in the design of protection schemes. No attempt will be made to summarise this document here.

The lightning protection system may comprise down conductors terminating in earth electrodes or may employ the building structural steelwork, with or without additional electrodes, but the combined earth termination network must not exceed 10 Ω resistance.

It is now a requirement of BS 6651[3], BS 7430[2] and BS 7671[4] that the lightning protection network be adequately bonded to the electrical system. This is a fairly recent change of policy as in former times the lightning protection system and the electrical earth system were carefully segregated.

BS 7671 limits the size of the bonding conductor to a size not exceeding that of the earthing conductor.

10.5.5   Remote installations

On many sites electrical power supplies are taken to buildings remote from the main intake. It is considered that the best practice is to provide a substantial CPC with the power cable to maintain a low impedance earth fault loop. An alternative is to regard the remote building as a separate consumer and provide a relatively high impedance CPC, i.e. the cable armouring. This may not permit sufficient fault current to flow and in this event a separate earth electrode would be required. It should be noted however that the neutral conductor must remain isolated from the protective conductor and the remote earth electrode.

10.5.6   Generator plant

Generator plant may be provided within an installation to operate in parallel with the public supply or in place of it in the event of a supply interruption, for example. The electricity supplier will usually agree to short periods (i.e. less than one minute) of parallel operation, even for standby plant.

Regardless of the form of electricity supply, if generators are to be incorporated an independent earth electrode is required. The machine framework and all exposed and extraneous conductive parts of the generating plant must be effectively connected to the main earthing terminal.

Neutral earthing requirements will vary depending on the nature of the public electricity supplier's (PES) (regional electricity company (REC)) supply, i.e. whether or not it is a TN-C (PME) system (see 10.3.3) and on the proposed method of operation.

10.5.6.1   Parallel (embedded) operation, non-PME supply

For this mode of operation the neutral point of the machine windings should not be earthed except in situations where the generating plant operates independently of the public supply. A changeover system is required so that the neutral point of the generator(s) is isolated while in parallel operation but connected to earth when operating alone, see Figure 10.14(a).

10.5.6.2   Parallel (embedded) operation, PME supply

With a PME supply the generator star point may be connected permanently to the supply company's PEN conductor at the main earthing terminal, see Figure 10.14(b). However if the generator is to be operated in parallel with a mains supply derived from the consumer's

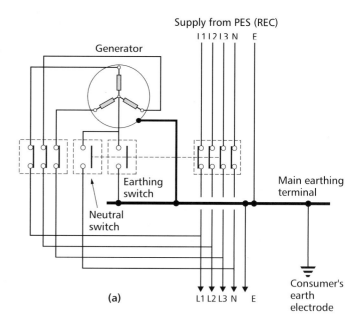

(a)

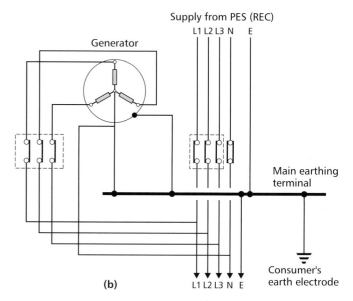

(b)

Figure 10.14 Generator earthing, parallel (embedded) operation; (a) non-PME supply, (b) PME supply

HV/LV distribution transformer it must be treated as a non-PME system.

10.5.6.3   Standby (island) operation

The star point should be permanently connected to the main earthing terminal for both supply systems. However it should be noted that a neutral switch is required for a non-PME supply, see Figure 10.15. It is important to note that where the LV supply is derived from a consumer's substation the neutral switch is not needed.

10.5.6.4   Multiple machines

Where it is anticipated that circulating currents, particularly third harmonic currents, could cause problems by virtue of their magnitude the following practices are available:

—   provision of a neutral earthing transformer enabling the neutral of the installation to be permanently earthed and the generator neutral points isolated

— star point switching to isolate all but one neutral point

— a tuned reactor in each generator neutral earth connection to attenuate higher frequency currents.

Modern low voltage generators have harmonic voltage levels sufficiently low not to require the above methods and it is quite common practice to provide permanent neutral earth connections for each machine.

10.5.6.5 High voltage generators

Impedance earthing is often applied to HV generators because the very high line-to-earth faults could otherwise damage the machine. Circulating currents are also more likely to prove harmful and an isolation method as described above will almost certainly be required.

10.5.6.6 Earth fault currents

Steady state earth fault currents for low voltage machines are generally in the order of:

— 1.5 to 2 times rated generator current without field forcing

— 2.5 to 3 times rated generator current with automatic voltage regulation (AVR) action.

For high voltage machines the value of earthing resistance or reactance must be chosen to produce earth fault currents consistent with the protection system. The range would normally be 25% to 100% of rated current.

10.6 Testing

10.6.1 Initial testing

Initial testing should include the following quantitative measurements:

— earth electrode resistance

— earthing conductor resistance

— earth loop impedance

— earth fault protection settings.

Accuracy expectations for each electrode resistance using the method shown in Figure 10.16 do not exceed 5%. This is perfectly acceptable as the system will vary seasonally.

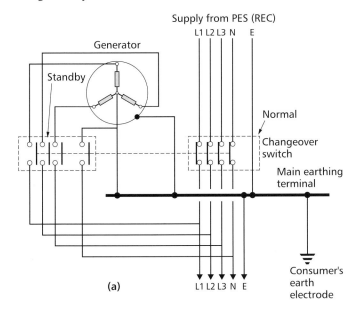

(a)

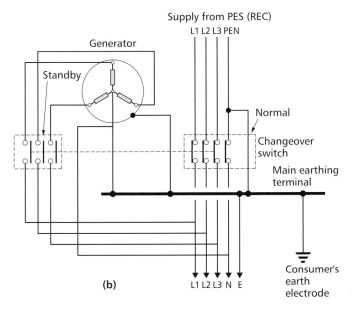

(b)

Figure 10.15 Generator earthing, standby (island) operation; (a) non-PME supply, (b) PME supply

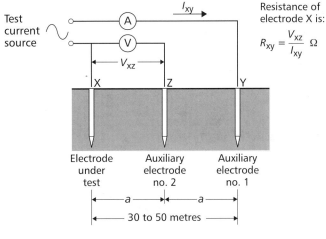

(a)

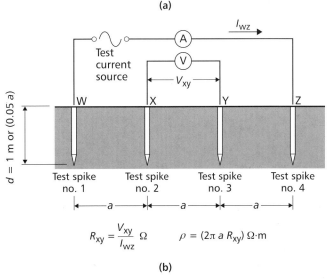

(b)

Figure 10.16 Resistivity testing; (a) electrode resistivity testing, (b) soil resistivity testing

Since it is not permitted to disconnect the electrodes once connected, it should be appreciated that the initial test will be the most thorough. Periodic testing is only possible with multiple electrodes by disconnecting individuals for testing. Complex calculations are then required to assess the system resistance.

The earth conductor resistance may be measured by either a direct reading DC ohmmeter or an AC isolated test set conforming to BS EN 60742 (BS 3535)[13]. The AC tester can produce high test currents up to about 25 A. The resistance of test leads and connections should obviously be kept as low as possible.

Proprietary earth loop impedance testers are usually of the AC mains type and work by drawing a current from the phase conductor and injecting this at low voltage back to the source via the earth return path.

Indicative testers, bells, flashers etc. must not be used as these provide no quantitative results.

10.6.2   Periodic testing

As mentioned above, periodic testing of earth electrode resistance is not very practicable. It is recommended that careful scrutiny only is carried out on this element. The same comments could apply equally to the earthing conductor.

Quantitative earth loop impedance tests and secondary injection testing of protection settings should be carried out at intervals not exceeding five years.

Testing of residual current devices (RCDs) with trip settings of 100 mA or less should be tested annually with a proprietary tester giving current and time readout. The test button on RCDs is not a reliable indicator of correct operation within declared limits.

Routine visual inspections should be carried out at least annually on all aspects of the earthing installation including main and supplementary equipotential bonding.

References

1   The Electricity at Work Regulations 1989 Stautory Instruments 1989 No. 635 (London: Her Majesty's Staionery Office) (1989)

2   BS 7430: 1998: *Code of practice for earthing* (London: British Standards Institution) (1998)

3   BS 6651: 1999: *Code of practice for protection of structures against lightning* (London: British Standards Institution) (1999)

4   BS 7671: 2001: *Requirements for electrical installations. IEE Wiring Regulations. Sixteenth Edition* (London: British Standards Institution) (2001)

5   BS 7361: Part 1: 1991: *Cathodic protection. Code of practice for land and marine applications* (London: British Standards Institution) (1991)

6   *Distribution Code of the Public Electricity Suppliers of England and Wales* (London: Office of Gas and Electricity Markets) (1990)

7   *Engineering recommendations for the connection of embedded generating plant to the RECs' distribution systems* Engineering Recommendation G.59/1 (London: Electricity Association) (1995)

8   *Notes for guidance for the protection of embedded generating plant up to 5 MW for operation in parallel with PES distribution networks* Engineering Technical Report ETR113 (London: Electricty Association) (1995)

9   *Protection Against Overcurrent* Guidance Note 6 to IEE Wiring Regulations (London: Institution of Electrical Engineers) (1999) ISBN: 0852969597

10   IEC 60364: *Electrical installations of buildings*: Part 1: 2001: *Fundamental principles, assessment of general characteristics, definitions* (Geneva: International Electrotechnical Commission) (2001)

11   BS 6701: 1994: *Code of practice for installation of apparatus intended for connection to certain telecommunication systems* (London: British Standards Institution) (1994)

12   BS EN 60309-2: 1998: (IEC 60309-2: 1997): *Plugs, socket-outlets and couplers for industrial purposes. Dimensional interchangeability requirements for pin and contact-tube accessories* (London: British Standards Institution) (1998)

13   BS EN 60742:1996: (BS 3535-1:1996): *Isolating transformers and safety isolating transformers. Requirements* (London: British Standards Institution) (1996)

11 Electromagnetic compatibility

11.1 Introduction

The world is bathed in electromagnetic waves. The frequency spectrum ranges from very high frequency, e.g. light, infrared, radio and TV waves, to very low frequency, e.g. 50 Hz power and below, plus direct current (DC). Much electrical equipment relies on receiving and transmitting electromagnetic radiation. The sensitivity of such equipment to extraneous radiation and the amount of ambient radiation must be compatible. Furthermore, there are statutory requirements to avoid causing interference to radio and TV systems.

In every aspect of modern living, electronics systems feature to some degree or other. Since, by definition, electronics must respond to electromagnetic energy, it is important to minimise the response to extraneous signals and to control or eliminate the coupling of electromagnetic interference to the electronic circuits and components. This is achieved through ensuring 'electromagnetic compatibility' (EMC), which is the ability of an electronic component, unit or system to co-exist with its electromagnetic (EM) environment.

As electronics finds wider and wider usage in a diversity of applications, the importance of ensuring that EMC is achieved will increase. Similarly, there will be an increase in the requirement to understand and predict all electromagnetic environmental effects. Today, with the proliferation of digital and communications equipment constantly finding new applications, the control of both external and internal electromagnetic interference (EMI) must be considered in any new building project. This is particularly true when considering the information technology of the 'electronic office', building management systems (BMS) and fire command systems.

It is important that all these systems operate correctly and consistently if the quality of life is not to be affected. If a TV remote control suffered from interference it would be a nuisance to a single individual. However, if the central record database of a large tax office or bank were corrupted it would affect a large sector of the population and would require considerable effort and cost to correct. Similarly, if certain types of process control were corrupted by interference a safety hazard could occur leading perhaps to a large-scale industrial accident.

In the design and construction of a building or complex there are two distinct EMC problems for the architect and the building services engineers to consider. On the one hand, a benign EM environment must be created by the building for the prospective client's personnel and equipment. This user environment must be provided over the predicted life of the building and may, therefore, have to account for future increases in the external EM pollution levels anticipated at the site. On the other hand, additional to the client's requirements, the building services engineer must consider the EM environment required for the successful operation of the building's management systems, information networks and process/plant control equipment. The latter may be more difficult to achieve since such equipment will be either located with, or at least connected to, the heavy plant of the building services.

11.2 EMC Directive and UK legal obligations

In order to establish free trade in electrical goods which are sensitive to electromagnetic radiation, the EU has issued Directives to control the sensitivity and the emissions of electromagnetic radiation by equipment on the market, and systems which are installed. The Directive[1] requires member states to enforce the requirements of the Directive. The UK legislation to implement the Directive is the Electromagnetic Compatibility Regulations 1992[2].

The EU directive has been explained in European Commission Guidelines[3].

This section of the Guide is intended to assist electrical engineers to provide electrical building services installations that comply with the requirements of this legislation.

Equipment which is supplied to the building should be sufficiently insensitive to operate successfully in the building environment; it should not emit excessive radiation. Broadly this means it should be manufactured to levels given in standards that implement the requirements of the Directive. Problems from electromagnetic interference (EMI) will be minimised by procurement of equipment that complies with relevant standards, is supplied with a relevant EMC declaration of conformity (DOC), and is installed and maintained using good EMC practices.

The installation must be designed and installed so that it does not generate electromagnetic radiation that is likely to cause interference to other systems that have the correct immunity for the intended environment.

11.2.1 The EMC Directive

The Directive[1] has proven to be very contentious for many reasons, caused partly and inherently by the fact it is a new-approach Directive. As such, the EMC Directive lays down essential protection requirements in broad terms, and leaves the task of defining specific technical requirements to the European standards-making bodies such as CENELEC and ETSI.

11.2.1.1   Aim of the EMC Directive

The EMC Directive has two inter-related aims as follows:

— Removal of barriers to trade between companies operating within the European Economic Area (EEA). This can only be achieved through the harmonisation of local legislation, approval procedures, performance standards and certification requirements.

— Achievement of an electromagnetic environment that is controlled through the application of essential protection requirements, allowing apparatus placed within that environment to operate as intended.

The essential protection requirements (Article 4) of the Directive are as follows:

The apparatus shall be so constructed that:

(a) The electromagnetic disturbance it generates does not exceed a level allowing radio and telecommunications equipment and other apparatus to operate as intended;

(b) The apparatus has an adequate level of intrinsic immunity to electromagnetic disturbance to enable it to operate as intended.

Both objectives of the Directive have clear benefits to manufacturers wishing to market and trade within the EEA.

In order to be placed onto the European market, the apparatus must comply with harmonised standards, reference to which has been published in the *Official Journal of the European Communities*. In addition, the conformity of the apparatus with this Directive must be certified by an EC declaration of conformity (DOC) issued by the manufacturer or its authorised representative established within the Community. The apparatus shall display the EC conformity mark, consisting of the letters 'CE'.

11.2.1.2   EC guidelines on the EMC Directive

The need for interpretation of the Directive has been addressed by the European Commission through the publication of amplified guidelines, the most recent version of which were published in July 1997[3]. These guidelines offer help on a range of issues including:

— Exclusions from the Directive, e.g. equipment that is covered by other Directives such as medical equipment.

— Definitions of and applications of the EMC Directive to components, finished products, systems and installations within the EMC Directive: building services engineers need to know the difference between equipment that is stand alone, e.g. an air conditioner operating in a single room, and multiple equipment systems and how these should be addressed when they are built into an installation.

— Responsibilities of manufacturers, importers, suppliers, system assemblers and installers: these responsibilities may include, for example, specifying compliant equipment, defining a compliant installation, and certifying compliance.

— Responsibilities of manufacturers and installers for the provision of management rules for the installation and continuing operation of systems in a building, with respect to containing electromagnetic emissions and ensuring that systems do not become more susceptible over time.

This code gives guidance to building services engineers to assist them in designing for electromagnetic compatibility. Basically, as a fixed installation, a building does not need to be CE-marked but it should not incorporate equipment which emits too much radiation or which is unduly sensitive.

An EMC management code of practice should give guidance on the application of the EMC Directive with respect to the installation and use of products and systems. A code of practice helps to ensure that the essential protection requirements of the EMC Directive are met or, at least, that due diligence has been adequately demonstrated when the systems in a building commence operation, i.e. the building is occupied.

11.2.2   The Electromagnetic Compatibility Regulations

The Electromagnetic Compatibility (EMC) Regulations[2] enact the requirements of the EMC Directive for the UK. Where a Directive, which has regulatory force, is introduced by the Commission, each European Member State must introduce its own legal regulations to implement the Directive. The regulation in each Member State should be identical. This may not always be the case but compliance with the UK regulations allows a UK manufacturer to place a product onto the market of other Member States. Complaints of non-compliance by another Member State should be addressed to the UK authorities, at the Department of Trade and Industry (DTI).

In principle, the UK regulations are divided into four broad categories:

— application of the regulations to relevant apparatus, and exclusions

— general requirements of supply and taking into service of apparatus

— routes to compliance via the use of standards, and the use of the technical construction file and EC-type examination route

— enforcement officers and procedures, including procedures and penalties when non-conforming apparatus is discovered.

Important terms within the regulations are defined in the following sections.

11.2.2.1   Relevant apparatus

The Directive applies to:

— electrical equipment

— systems made up from combinations of electrical equipment.

Relevant apparatus is electrical apparatus for which the majority of the regulations apply, and is defined as:

— a product with an intrinsic function intended for the end user; and

— is supplied or intended for supply or taken into service or intended to be taken into service, as a single commercial unit, which is:

  (a)   an electrical appliance

  (b)   an electronic appliance

  (c)   a system.

It is important to note that relevant apparatus includes equipment and systems that are purchased in a number of ways. Examples are as follows:

— 'off the shelf' equipment, e.g. a TV set

— 'customised' equipment, e.g. a pumping set

— 'built to order' equipment, e.g. a control panel.

The designer and manufacturer of the relevant apparatus is the person responsible for complying with the EMC Directive. The Directive requires that a single person within an organisation be authorised to bind the company in meeting the requirements of the Directive.

11.2.2.2   Installation

The Directive also applies to installations.

An installation, within the terms of the EMC Directive, can be either fixed or movable. A fixed installation is defined as[3]:

> a combination of several equipment, systems, finished products and/or components (hereinafter called parts), assembled and/or erected by an assembler/installer at a given place to operate together in an expected environment to perform a specific task, but not intended to be placed on the market as a single functional or commercial unit.

This type of installation must comply with the essential requirements of the EMC Directive but, as it cannot enjoy free movement within the EU market, there is no need for CE marking or an EC declaration of conformity (DOC). Yet, the person responsible for the design, engineering and construction of the installation becomes the manufacturer in the sense of the Directive. That person then assumes the responsibility for the installation's compliance with all applicable provisions of the Directive.

A movable installation must comply with all provisions of the Directive, like a system as described above. However, some movable installations are intended to replace or extend a fixed installation and should be treated as fixed installations, if their connection to an existing fixed installation is carefully planned and set up.

A moveable installation such as a TV outside broadcast vehicle or a prefabricated building should be marked so that it can travel freely throughout Europe.

11.2.2.3   Supply and taking into service

From the UK Regulations[2], regulations 28 and 30 require that no person shall supply relevant apparatus unless it:

— conforms with the protection requirements

— meets the conformity assessment requirements

— has the CE marking properly applied

— has an EC declaration of conformity certificate.

Regulation 29 requires that no person shall take into service relevant apparatus unless it conforms with the protection requirements.

For example, apparatus covered by the EMC Directive is taken into service when it is first used, i.e. by the end user that operates the apparatus, e.g. escalators.

'Taking into service' does not include the area of energising, testing and commissioning of the apparatus by the manufacturer before the handover of the apparatus to the end user.

The apparatus manufacturer will be in a position of overall control in ensuring that the essential protection requirements are satisfied and assumes legal responsibility for compliance. Data must also be provided for the end user to ensure that these requirements are satisfied throughout the operational life of the apparatus.

11.2.2.4   Presumption of conformity

Regulation 32 states that relevant apparatus supplied with the correct EMC certification and conforming to the appropriate standards or technical construction file declaration can be presumed to conform to the protection requirements.

This presumption remains valid if the relevant apparatus supplied is installed, maintained and used for the purposes for which it was intended.

This presumption means that the operator or owner of the building or installation is not required to carry out (or have carried out) any further tests if supplied components and equipment are operated within the specified environment.

11.2.2.5   CE mark

All apparatus covered by the Directive in accordance with the protection requirements and accompanied by certification documents such as a declaration of conformity (DOC) must bear the CE mark. Those who specify equipment for installation in buildings must ensure that CE-marked equipment is satisfactory for use within that building, e.g. the requirements for an air conditioning unit's EMC specification will vary depending on whether it is to be used in a domestic or a heavy industrial environment

11.2.2.6   User manual

All equipment, systems or installation should be provided with a user manual for operation and maintenance. This manual should include essential information on EMC, so that the user is aware of the intended use and limitations of the apparatus, and how EMC should be dealt with in order to ensure the good operation of the apparatus during its whole life-cycle.

The instructions for use must contain all the information required in order to use the apparatus in accordance with the intended purpose and in the defined electromagnetic environment. These instructions must give the following information:

— intended conditions of use

— instructions on:

(a) installation

(b) assembly

(c) adjustment

(d) taking into service

(e) use

(f) maintenance

(g) where necessary, warnings about limitations on use.

11.3 Responsibilities for EMC compliance

The building services engineer and the owner/operator of a building each have responsibility, under the Directive to ensure that compliance with the directive is met and maintained. These responsibilities are outlined below.

11.3.1 Responsibilities of the building services engineer

The person(s) responsible for the design, engineering and construction (assembly and erection) become(s) the 'manufacturer' for the purposes of the EMC Directive, and assume(s) the responsibility for the installation's compliance with all applicable provisions of the Directive, when taken into service. The whole method of installation has to be in accordance with good engineering practice within the context of installation, as well as installation rules (local, regional, national) that will ensure the compliance of the whole installation with the essential requirements of the EMC Directive.

11.3.2 Responsibilities of the installation operator

The operator is the person who uses the installation, possibly daily. The operator can reasonably expect to be provided with the maximum of information about the installation. The information needed may include:

— knowledge of the exact purpose of each part of the installation

— knowledge of the EMC environment

— in the case of a public installation (e.g. a shopping mall or a railway station), particular data for care to be taken to prevent any safety risk

— knowledge of who is responsible for the maintenance of the installation.

The same questions arise for modifications to the installation.

In other words, the operator needs very comprehensive instructions for the installation, with very clear explanations concerning the EMC environment and management, see section of 11.2.2.6.

11.4 EMC requirements as part of the project brief

This section refers to the definition of the electromagnetic environment that should be considered for the project brief specification, together with the various standards which need to be considered, if relevant to the project.

11.4.1 Evaluation of the electro-magnetic environment

The project brief defines specifications that should be implemented by the contractor in order to provide a safe and reliable installation. Among the specifications, some are related to the EMC aspect of an installation. The electromagnetic environment is an important aspect to consider in the design of a new installation. It will indeed put constraints on:

— the equipment to purchase, which will need to comply to different sets of specifications

— the equipment layout

— the necessary level of protection for each item of equipment or system.

Some standards define electromagnetic environments. These are outlined in the following sections.

11.4.1.1 BS EN 61000-6-1 and BS EN 61000-6-3

BS EN 61000-6-1[4] and BS EN 61000-6-3[5] define the domestic, commercial and light industry environment. Clause 5 of BS EN 61000-6-3 reads:

The environments encompassed .... are both indoor and outdoor. The following list, although not comprehensive, gives an indication of locations which are included:

— Residential properties, e.g. houses, apartments, etc;

— Commercial premises e.g.

— Retail outlets, e.g. shops, supermarkets, etc;

— Business premises, e.g. offices, banks, etc;

— Areas of public entertainment, e.g. cinemas, public bars, dance halls, etc;

— Outdoor locations, e.g. petrol stations, car parks, amusement and sports centres, etc;

— Light-industrial locations, e.g. workshops, laboratories, service centres, etc.

Locations which are characterised by being supplied directly at low voltage from the public mains are considered to be residential, commercial or light industrial.

Radiated emission limits are given at a measurement distance of 10 metres.

11.4.1.2   BS EN 61000-6-2 and BS EN 61000-6-4

BS EN 61000-6-2[6] and 61000-6-4[7] define the industrial environment. Clause 5 of BS EN 61000-6-4 reads:

Industrial locations are characterised by the existence of one or more of the following conditions:

— Industrial, scientific and medical apparatus (ISM) are present;

— Heavy inductive or capacitive loads are frequently switched;

— Currents and associated magnetic fields are high.

— These are the major contributors to the industrial electromagnetic environment and as such distinguish the industrial from the other environments.

The scope of this standard adds:

Apparatus covered by this standard is not intended for connection to a public mains network but is intended to be connected to a power network supplied from a high- or medium-voltage transformer dedicated for the supply of an installation feeding manufacturing or similar plant. This standard applies to apparatus intended to operate in industrial locations or in proximity of industrial power installations.

11.4.1.3   IEC 61000-2-5

Standard IEC 61000-2-5[8] in particular provides a rationale for assessing the relevant type of installation, and therefore the levels likely to be encountered, as follows:

— Eight types of installations are defined, according to the types of equipment that can be encountered within the enclosure (or at a close proximity), the type of AC and DC power supply, the signal/control lines and earthing practices.

— For these types of installation, and for each type of disturbance, there are associated disturbance categories are associated, which are defined as follows:

   (a)   Class A: controlled environment, i.e. controlled by the nature of the building or any particular installation practices.

   (b)   Classes 1, 2, etc. (up to 5, depending on the type of disturbance considered) correspond to natural environments, and are classified from the lowest to the highest levels.

   (c)   Class X (harsh environment) is an environment where the degree of disturbance is higher than is generally encountered.

   The levels for classes A and X are defined on a case-by-case basis.

— For each type of disturbance, tables are provided with the level likely to be encountered according to the degree of disturbance.

The correct starting point is to use IEC 61000-2-5, which provides an overall description of all types of installations and then likely levels encountered for many types of disturbances. This then defines the type of equipment to be purchased.

11.4.2   EMC standards for specific equipment and cables

Many standards have been issued that cover equipment and cables. Some of these are given below, a comprehensive list can be viewed on the European Commission's website*.

11.4.2.1   Equipment

Buildings contain different kinds of equipment, all of which must comply with relevant EMC standards. Examples[4-7,9-17] of these are presented in Table 11.1. Current standards can be viewed on the EC website†.

11.4.2.2   Cables

It is also possible to find low voltage and high voltage cables run together with telecommunications and IT cables. Therefore, it is necessary to use the relevant standards during the various stages of a new building construction. These are mainly from the BS EN 50174[18]

---

* http://europa.eu.int/comm/enterprise/newapproach/standardization/harmstds/reflist.html

† http://europa.eu.int/comm/enterprise/newapproach/standardization/harmstds/reflist/emc.html

Table 11.1 EMC standards for equipment

Type of equipment	Applicable standard(s)
Access control units	BS EN 50130-4[9] plus Admt. 1: 1998
Air handling units	BS EN 61000-6-1/2/3/4[4-7]
Audio amplifiers	BS EN 61000-6-1/2/3/4[4-7]
Battery charger	BS EN 61000-6-1/2/3/4[4-7]
Boilers	BS EN 61000-6-1/2/3/4[4-7]
CCTV control panels	BS EN 61000-6-1/2/3/4[4-7]
Chillers	BS EN 61000-6-1/2/3/4[4-7]
UPS	BS EN 50091-2[10]: 1995
Extract fans	BS EN 61000-6-1/2/3/4[4-7]
Fire alarm system	BS EN 50130-4[9] Admt. 1: 1998 BS EN 50270[11]
HV switch gear	BS EN 61000-6-1/2/3/4[4-7]
HVAC control system	BS EN 60730-2[12]
IT equipment used in BMS/EMS, CCTV, access control, intruder alarm, and fire detection systems	BS EN 55022[13]
Lifts	BS EN 12015[14] (emission) BS EN 12016[15] (immunity)
Lighting equipment	BS EN 55015[16] plus Admt. 1: 1997 and Amdt. 2: 1999
LV switch gear	BS EN 60947[17]
Power distribution units	BS EN 61000-6-1/2/3/4[4-7]

Table 11.2 Standards for cabling

Phase	Relevant standard(s)	Subject areas
Building design phase	BS EN 50310[19]	Common bonding network (CBN) within a building; AC distribution system and bonding of the protective conductor (TN-S)
Cabling design phase	BS EN 50173[20] or (and)	
	BS EN 50098-1[21] or (and)	
	BS EN 50098-2[21] or (and) other application standards	
Planning phase	BS EN 50174-1[18] and	Specification considerations; quality assurance; cabling administration
	BS EN 50174-2[18] and	Safety requirements; general installation practices for metallic and optical fibre cabling additional installation practice for metallic cabling; additional installation practice for optical fibre cabling
	BS EN 50174-3[18] and	Equipotential bonding
	BS EN 50310[19]	Common bonding network (CBN) within a building; AC distribution system and bonding of the protective conductor (TN-S)
Implementation phase	BS EN 50174-1[18] and	Documentation; cabling administration
	BS EN 50174-2[18] and	Safety requirements; general installation practices for metallic and optical fibre cabling; additional installation practice for metallic cabling; additional installation practice for optical fibre cabling
	BS EN 50174-3[18] and	Equipotential bonding
	BS EN 50310[19]	Common bonding network (CBN) within a building; AC distribution system and bonding of the protective conductor (TN-S)
Operational phase	BS EN 50174-1[18]	Quality assurance; cabling administration; repair and maintenance

series of standards. These and other relevant standards[19–21] are summarised in the Table 11.2. Different standards may be used at different stages and this is illustrated in the table.

11.5 Construction management plans

With the increasing integration of electrical with electronic systems, there is a greater need to successfully manage the EMC compliance of the whole building and its infrastructure. Certain rules need to be implemented; some of these are described below.

EMC compliance for building management systems (BMS) requires a total EMC approach to installing systems such that compliance becomes an integral part of the system's life-cycle. Consequently, continuing compliance involves a variety of business disciplines not normally considered to be involved in the EMC process. The business framework requires that EMC codes of practice and quality assurance systems are introduced at all stages of design, development, approval, procurement, installation and operation. In addition, procedures should be in place, which address ongoing maintenance up to the withdrawal of systems in use.

The starting point for achieving whole-life EMC management is the preparation of an EMC management control plan. The objective of such a plan is to set out what needs to be done so that systems achieve compliance from completion of the installation, continue to comply during modifications and up to replacement by more modern systems. All of this should be achieved at the lowest possible cost. In addition, the EMC control plan should be part of the existing business procedures (e.g. quality assurance, quality control) and not stand outside of them.

Some of the contents of a typical management control plan are outlined as follows:

— product definition, in terms of the Directive

— electromagnetic environment assessments and risk analysis

— definition of legal requirements, compliance routes and decision flow charts

— purchasing rules and supplier audits

— conformity in installation

— installation, assembly and operation techniques.

11.6 Good design practices

Good EMC design practices applied at the commencement of design will significantly reduce the installation and remedial costs inherent in not applying the practices until later in the project. Best practices that have the most significant effect on containing electromagnetic problems are given below.

11.6.1 System shielding

Electromagnetic disturbances can be either radiated or conducted. It is therefore essential to protect the environment from such disturbances (emission problem) and/or to be protected from the disturbances in the environment (immunity problem). Both radiated and conducted disturbances must be taken into account. There are several general approaches to obtain EMC immunity for a system, these are shown in Figure 11.1 and are described below:

(a)   *Global protection by single barrier*: a screen surrounds all devices of a system, and suitable devices protect

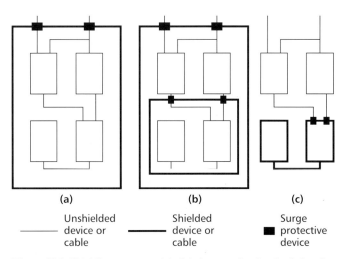

Figure 11.1 Shielding strategy; (a) global protection by single barrier, (b) global protection by multiple barriers, (c) distributed protection

all cable entries. No specific protection is applied to the individual units, see Figure 11.1(a).

(b) *Global protection by multiple barriers*: same as above, but some particularly sensitive or noisy devices are separated in a second enclosure, nested in the first one. There can be as many zones nested in one another as necessary, see Figure 11.1(b).

(c) *Distributed protection*: only the sensitive and/or noisy devices are protected, in an individual manner, see Figure 11.1(c).

In all situations, there is a separation between electromagnetically noisy zones and quiet zones, and there may be several levels of quiet zones, as in case (b). These separations are an essential process in the shielding strategy, since it is not often possible to attenuate disturbances by 80 dB at all frequencies with the use of a single screen. Moreover, some systems do not need such attenuation and providing a high level of shielding effectiveness is often expensive. Therefore, a cost and risk analysis should be performed in order to choose the most appropriate strategy.

The shielding effectiveness (S_{dB}) is obtained by the product of three terms, each representing one loss parameter, i.e. reflection loss (R), absorption loss (A) and multiple reflection (M). The situation is indicated in Figure 11.2. Expressing the factors in dB terms, they may be added to give:

$$S_{dB} = A_{dB} + R_{dB} + M_{dB} \qquad (11.1)$$

where S_{dB} is the shielding effectiveness (dB), *AdB is the* absorption loss (dB), R_{dB} is the reflection loss (dB) and M_{dB} is the multiple reflection (dB).

The internal multiple reflections (M_{dB}) are negligible when the absorption loss (A_{dB}) is greater than 10 dB. The absorption loss depends mainly on the characteristics of the material and is defined by:

$$A_{dB} = 131.66\, t \sqrt{f\, \mu_r\, \sigma_r} \qquad (11.2)$$

where t is the thickness (m), f is the frequency (MHz), μ_r is the permeability (H·m^{-1}) and σ_r is the conductivity (S·m^{-1}).

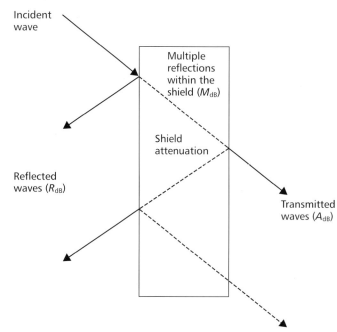

Figure 11.2 Illustration of shielding effectiveness

The reflection loss (R_{dB}) is caused by the impedance mismatch between the incident wave and the shield. It can be shown that the source wave impedance depends on the type of the source (i.e. electric or magnetic) and the distance from the source. This gives three relations for R_{dB}, as follows.

For planes waves:

$$R_{dB} = 168 - 10 \log (f\, \mu_r\, /\, \sigma_r) \qquad (11.3)$$

For an electric field source:

$$R_{dB} = 168 - 10 \log (r^3 f^3\, \mu_r\, /\, \sigma_r) \qquad (11.4)$$

For a magnetic field source:

$$R_{dB} = 168 - 10 \log (\mu_r^2\, /\, r^2 f\, \sigma_r) \qquad (11.5)$$

where r is the distance from the source (m).

A list of typical shield materials is given in Table 11.3, together with values of μ_r, σ_r, $\sqrt{(\mu_r\, \sigma_r)}$ and $[10 \log (\mu_r / \sigma_r)]$.

It can be seen that A_{dB} for iron is some 14 times that for copper, while R_{dB} is about 37 dB lower.

The amount of shielding to achieve the desired electromagnetic field inside a building can be calculated. However, the problem is complex, depending on the

Table 11.3 Characteristics of screening materials

Metal	μ_r	σ_r	$\sqrt{(\mu_r/\sigma_r)}$	$10 \log_{10} (\mu_r/\sigma_r)$
Copper	1	1	1	0
Aluminium	1	0.6	0.8	2.3
Magnesium	1	0.4	0.5	4
Zinc	1	0.3	0.4	5
Nickel	1	0.2	0.3	7
Lead	1	0.08	0.2	11
Iron	10^3	0.2	14	37
Steel	10^3	0.1	10	40

number of radio frequency sources outside and inside the building, the building construction materials and the amount of internal reflections caused by the type of building materials used. Specialist EMC advice should be sought that addresses both measurement and calculation of these fields.

11.6.2   Cable segregation and separation

11.6.2.1   Cable segregation

Cables should be segregated from each other depending on their function. In EMC engineering practice, the following general guidelines should be observed:

— Mains cables, including power feeds and lighting circuits, carrying up to 250 V should not be grouped with sensitive cables (i.e. data cables).

— All cabling should avoid any proximity to radio or television transmitters, beacons and overhead transmission lines.

— High-level impulses on cables, due to their fast rise times, produce a large frequency distribution of disturbance. Special precautions need to be taken where these types of cabling exist; heavy screening, clean earthing at both ends and an increase in the separation with adjacent cables would need to be implemented.

— All cables should be terminated in accordance with their characteristic impedance. The formation of the wires in the cable determine the resistance (impedance) of the cable between the wires. To achieve best transfer of energy (at 50 Hz and at higher frequencies used for communications, for example), requires that the connection to equipment (connector) should continue the impedance through into the equipment. If this is not done, then energy is lost as radiation into the environment. This spurious radiation (emission) is what causes interference into other systems.

— Metal cable trays, if not already in use, should be installed. With an adequate mesh for the frequencies in use and good earthing, the tray will effectively become a screen or enclosure for the cables. Multi-compartment trunking is usually used but cable separation as defined below must be maintained.

11.6.2.2   Cable separation

To reduce capacitive and inductive coupling between cables, they should be separated as far as possible from each other. In particular, it is necessary to separate power cables from digital cables. Tables 11.4 to 11.9 summarise the cable separation recommendation. They are given for different building installation environments.

11.6.3   System layout

Within an installation the layout of equipment is as important as the layout of cables. Like cables, the rule is to separate radiating equipment from sensitive equipment, so the first approach is to gather the equipment with high emission characteristics at one part of the installation. On a large site, this equipment may require to be located at a minimum distance of 30 m from residential premises, to prevent them from being affected by radiated interference. If possible, the same distance should be maintained between electrically noisy and sensitive equipment within the site. This may be difficult on a small site, so the solution is to isolate the equipment that could be a source of interference by enclosing it within either a shielded enclosure or a shielded building.

Table 11.4 Guide to separation tables

Type of installation	Coupled cables involved		Table number(s)
	Source	Victim	
Single occupancy IT building	Power	Data	11.5, 11.6, 11.7
Multi-occupancy IT building	Power	Data	11.5, 11.6, 11.7
	Power	Control	11.7
	3-phase	Control	11.7
Heavy industrial environment	Power	Data	11.5, 11.6, 11.7
	Power	Control	11.7
	3-phase	Control	11.7
	Switched power line	Control	11.7
Hospital environment	Power	Data	11.5, 11.6, 11.7
	Power	Control	11.7
	3-phase	Control	11.7
Railways over or under-ground	Switched power line	Data	11.7
	Switched power line	Control	11.7
	Switched power line	Data (crossing)	11.9
Shopping mall	Power	Data	11.5, 11.6, 11.7
	3-phase	Data	11.7
	Switched power line	Data	11.7
Common services	Switched power line	Control	11.7
	Fluorescent lamps	Data	11.7
Any type of installation	Parallel runs of cables		11.8

Table 11.5 Separation between power and signal cables according to shielding

Signal cable type	Separation from stated power cable type / mm		
	Twin and earth	Steel wire armoured	MICC
Plain	150	125	Touching
UTP	75*	50	Touching
	125†	50	Touching
Screened	Touching	Touching	Touching

* below 100 MHz
† above 100 MHz

Table 11.6 Separation between power and signal cables according to the method of separation

Cable type	Separation for stated separation method / mm	
	Metal	Plastic
Unscreened power lines, or electrical equipment and unscreened IT lines	150	300
Unscreened power lines, or electrical equipment and screened IT lines *	30	70
Screened power lines*, or electrical equipment and unscreened IT lines	2	3
Screened power lines*, or electrical equipment and screened IT lines*	1	2

* 'screened' = screened and earthed at both ends

Table 11.7 Separation between source and victim cables when unshielded

Susceptibility of victim cable	Separation from source cable of stated noise level / mm			
	Indifferent	Noisy	Very noisy	Fluorescent lamp
Very sensitive	300	450	600	—
Sensitive	150	300	450	130

Table 11.8 Maximum parallel runs of cables

Type of cable	Separation / mm	Maximum parallel run / m
Any; very well armoured and separated by a metallic divider	Close	100
IT/power	25	5
	50	9
IT/power (under carpet)	130	15
	460	>15

Table 11.9 Minimum clearances between overhead telecommunications and power lines

Telecom. line	Minimum clearance from power line carrying stated voltage /m			
	< 1000 V		> 1000 V	
	Conductors	Poles	Conductors*	Poles
Cables	0.5	—	$(3 + 0.0015\,U)$	—
Poles	0.5	0.3	$(1.5 + 0.0015\,U)$	>1

* U = voltage in kV

11.7 Hazards of non-ionising radiation

For some years there has been concern regarding limits for safe exposure from the non-ionising radiation emitted from equipment such as transformers and cables. Standards have been published that set limits both nationally and internationally.

In April 1998, the International Commission on Non-Ionising Radiation Protection (ICNIRP) published guidelines[22] for limiting exposure to time-varying electric, magnetic and electromagnetic fields in the frequency range up to 300 GHz. These guidelines, which are intended to prevent the acute effects of exposure to DC, low frequency, radio frequency (RF) and microwave fields on humans, are based on biological considerations from which dosimetric quantities are derived, including a suitable safety margin. These are known as 'basic restrictions'. These quantities, relating to electromagnetic energy absorption and current flow in the tissues of the body, are not directly measurable. However, electromagnetic quantities such as electric and magnetic field strengths and the equivalent plane wave power flux density (PFD) may be derived from them, assuming whole body exposure in the far-field. These values are known as the 'reference levels' in the ICNIRP guidelines[22] and are given in Figures 11.3 and 11.4. The ICNIRP guidelines give reference levels for occupational and the general public exposures.

Within the UK, the National Radiological Protection Board (NRPB) has set limits for safe exposure. These limits are similar to those expressed in the ICNIRP guidelines, but expert guidance should be sought in order that proper levels are defined or measured.

11.7.1 ICNIRP basic restriction

The basic restrictions on the electromagnetic exposures are presented in Table 11.10.

11.7.2 ICNIRP reference levels

The reference levels for low frequency exposures (0 Hz to 100 kHz) are based on instantaneous short-term effects. These include surface charge perception effects, nerve stimulation and shocks from ungrounded large metallic structures and vehicles immersed in low frequency fields. Reference levels for high frequency exposures, 100 kHz to 300 GHz, are based on whole body heat stress, localised heating of sensitive body tissues and superficial heating, as well as radio-frequency burns resulting from contact with ungrounded large metallic structures and vehicles immersed in high frequency fields. All reference levels given by ICNIRP are presented in a two-tier system, one set of levels for occupational exposures and the other set for general public exposures. In general, the occupational exposure levels are a factor five greater than exposure levels for the general public, as shown in Table 11.11.

11.8 Effect of the external EM environment

The external environment must also be considered. For example, is the building going to be near to high power radar sources such as those found at airports? If safety critical equipment is to be installed in the building, e.g.

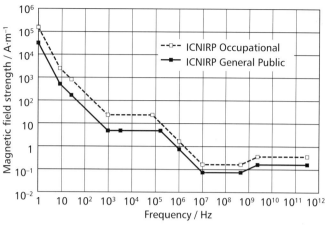

Figure 11.3 ICNIRP reference levels for magnetic field strength[22]

Figure 11.4 ICNIRP reference levels for electric field strength[22]

Table 11.10 ICNIRP basic restrictions for occupational and the general public for frequencies up to 300 GHz

Frequency range	Exposure	Basic restriction★	Comments★
Up to 1Hz	Occupational	40 mA·m⁻²	Induced current density in the head and trunk. All current density values given are RMS. For frequencies below 100 kHz the peak current density can be calculated by multiplying the RMS value by √2. Because of the electrical inhomogeneity of the body, the current density should be averaged over a section 1 cm², normal to the current flow. For frequencies up to 100 kHz and for pulsed magnetic fields, the maximum current density can be calculated from the rise/fall times, and the maximum rate of change of magnetic flux density. The induced current density can then be compared with the appropriate basic restriction.
	General public	8 mA·m⁻²	
1 Hz to 4 Hz	Occupational	$(40/f)$ mA·m⁻²	As above
	General public	$(8/f)$ mA·m⁻²	
4 Hz to 1 kHz	Occupational	10 mA·m⁻²	As above
	General public	2 mA·m⁻²	
1 kHz to 10 MHz	Occupational	$(f/100)$ mA·m⁻²	As above
	General public	$(f/500)$ mA·m⁻²	
100 kHz to 10 GHz	Occupational	0.4 W·kg⁻¹ 10 W·kg⁻¹ 20 W·kg⁻¹	Whole body average SAR Localised SAR in the head and trunk Localised SAR in the limbs
	General public	0.08 W·kg⁻¹ 2 W·kg⁻¹ 4 W·kg⁻¹	Whole body average SAR Localised SAR in the head and trunk Localised SAR in the limbs
10 GHz to 300 GHz	Occupational	50 W·m⁻²	Power density
	General public	10 W·m⁻²	

★ f = frequency (Hz); SAR = specific absorption rate

Notes:

(1) All SAR values are to be averaged over 6-minute period.

(2) Localised SAR averaging mass is 10 g of the exposed tissue and is taken in a volume of 1cm³; the maximum SAR so obtained should be the value used for the estimation of exposure.

(3) For pulses of duration t_p the equivalent frequency to apply in the basic restrictions should be calculated as: $f = 1/(2\,t_p)$

(4) To avoid auditory effects in the frequency range 0.3 GHz to 10 GHz from pulsed exposure, the SAR should not exceed 10 mJ·kg⁻¹ for workers and 2 mJ·kg⁻¹ for the general public, averaged over 10 g of tissue.

(5) Power densities are to be averaged over any 20 cm² of exposed area and any minute period to compensate the decreasing penetration depth with increasing frequency.

(6) Spatial maximum power densities, averaged over 1 cm², should not exceed 20 times the values of the power densities given.

life support systems in a hospital, then additional mitigation measures, such as building shielding or filtering of cables into the building, may need to be applied. Where the building is located over high magnetic or electric field sources, e.g. above an underground railway system, then analysis of the effects on systems installed in the building should be performed. A risk analysis as described below, will be required to be performed.

11.9   Effect of new technologies on maintaining compliance

The advanced electronic controls that are widely used in providing building services can be affected by electrical noise from sources such as IT equipment, industrial machinery and RF transmitters.

Table 11.11 ICNIRP reference levels for occupational and general public exposure from 0 Hz to 300 GHz

Occupancy	Frequency range	ICNIRP reference level			
		Electric field strength / V·m⁻¹	Magnetic flux density (B) / mT	Magnetic field strength (H) / A·m⁻¹	Equivalent plane wave PDF / W·m⁻²
Occupational	Up to 1 Hz	—	200 000	163 000	—
General public	Up to 1 Hz	—	40 000	32 000	—
Occupational	1 Hz to 8 Hz	20 000	$200\,000/f^2$	$163\,000/f^2$	—
General public	1 Hz to 8 Hz	10 000	$40\,000/f^2$	$32\,000/f^2$	—
Occupational	8 Hz to 25 Hz	20 000	$25\,000/f$	$20\,000/f$	—
General public	8 Hz to 25 Hz	10 000	$5\,000/f$	$4\,000/f$	—
Occupational	25 Hz to 820 Hz	$500/f$	$25/f$	$20/f$	—
General public	25 Hz to 800 Hz	$250/f$	$5/f$	$4/f$	—
Occupational	0.82 kHz to 65 kHz	610	30.7	24.4	—
	65 kHz to 1 MHz	610	$2.0/f$	$1.6/f$	—
General public	0.8 kHz to 3 kHz	$250/f$	6.25	5	—
	3 kHz to 150 kHz	87	6.25	5	—
	150 kHz to 1 MHz	87	$0.92/f$	$0.73/f$	—
Occupational	1 MHz to 10 MHz	$610/f$	$2.0/f$	$1.6/f$	—
General public	1 MHz to 10 MHz	$87/f^{1/2}$	$0.92/f$	$0.73/f$	—
Occupational	10 MHz to 400 MHz	61	0.2	0.16	10
General public	10 MHz to 400 MHz	28	0.092	0.073	2
Occupational	400 MHz to 2 GHz	$3f^{1/2}$	$0.01f^{1/2}$	$0.008f^{1/2}$	$f/40$
General public	400 MHz to 2 GHz	$1.375f^{1/2}$	$0.0046f^{1/2}$	$0.0037f^{1/2}$	$f/200$
Occupational	2 GHz to 300 GHz	137	0.45	0.36	50
General public	2 GHz to 300 GHz	61	0.2	0.16	10

Note: f = frequency (Hz)

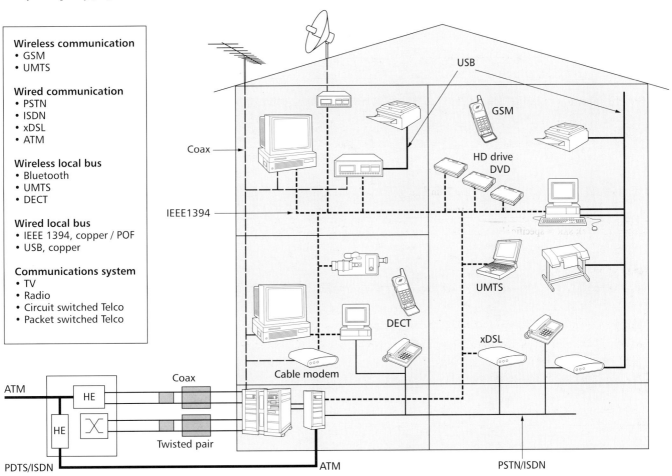

Figure 11.5 Typical data and control systems

The newer electronic systems, see Figure 11.5, offer a wide range of facilities but in certain cases they will be more susceptible to electrical interference. The level of electrical noise in the environment is increasing as the use of electrical and electronic equipment of all sorts is installed and used in buildings. Most of the sensors operate at low level and so are particularly sensitive to interference. The electromagnetic fields generated by radio systems and/or mobile telephones generate induced voltage currents and voltage on signal or control conductors. So they may have an impact on safety implications e.g. fire alarm controllers.

To analyse the potential EMC risk, a risk matrix can be used. An example of a this matrix is given as Table 11.12. It lists the likely sources (rows) and victims (columns) of disturbance in a building. 'Yes' indicates that there is potentially an EMC problem between a given source and a given victim, while 'No' means that no EMC problem should occur between the two systems.

Where a 'Yes' is given, then further analysis and investigation is required. An example of the analysis needed is given below.

Each intersection should be analysed and an explanation presented. For example, for row D:

— Row D ('Yes'): LV switchgear and distribution board are the source of the same type of disturbance. Nearly all systems are supplied by the LV system so that the disturbances are propagated to many types of systems. For those systems marked 'Yes', further analysis will be required.

— Row D ('No'): those systems marked with a 'No' are new systems which will not be connected to the existing LV switchgear so that no interference problems should occur.

This analysis is the first step of a complete study of EMC risk. The findings of this analysis show that a quantitative risk assessment should be performed.

References

1   Council Directive of 3 May 1989 on the approximation of the laws of the member states relating to electromagnetic compatibility EC Directive 89/336/EEC *Official J. of the European Communities* 23.5.89 L139/19 (as amended by Council Directives 92/31/EEC (1992), 93/68/EEC (1993) and 93/97/EEC (1993)) (Brussels: Commission of the European Communities) (1989–93)

2   The Electromagnetic Compatibility Regulations 1992 Statutory Instruments 1992 No. 2372; The Electrocompatibility (Amendment) Regulations 1994 Statutory Instruments 1994 No 3080; The Electrocompatibility (Amendment) Regulations 1995 Statutory Instruments 1995 No 3180 (London: Her Majesty's Stationery Office) (1992–95)

3   *Guidelines on the application of council directive 89/336/EEC* (Brussels: Commission of the European Communities) (1997) (with subsequent amendments)

4   BS EN 61000-6-1: 2001: *Electromagnetic compatibility (EMC). Generic standards. Immunity for residential, commercial and light-industrial environments* (London: British Standards Institution) (2001)

Table 11.12 Example risk matrix

Source system	Victim system							
	1	2	3	4	5	6	7	8
	CCTV	Telephone	HVAC plant	Fire alarms	Lifts and escalators	IT equipment	LV switchgear and power distribution boards	Control rooms
(A) 11 kV cables	No	No	No	No	Yes	Yes	No	No
(B) HV switchgear	No	No	No	No	No	Yes	Yes	No
(C) Transformers	No	No	Yes	No	Yes	Yes	Yes	No
(D) LV switchgear	Yes	Yes	No	Yes	No	Yes	—	No
(E) Small power	No	No	No	No	No	Yes	No	No
(F) Lighting	No	No	No	No	No	No	No	No
(G) CCTV	No	No	No	No	No	No	No	No
(H) Telephone	No	—	No	No	No	No	No	No
(I) HVAC plant	No	No	—	No	No	No	No	No
(J) Fire alarm	No	No	No	—	No	No	No	No
(K) Water pumps	No	No	No	No	No	No	No	No
(L) Lifts and escalators	No	No	No	No	—	No	No	No
(M) Mobile phones	No	Yes	No	No	Yes	Yes	No	Yes
(O) Broadcast transmitters	No	No	No	No	No	No	No	No
(P) ISM equipment	No	Yes	No	No	No	Yes	No	Yes
(Q) IT equipment	No	No	No	No	No	—	No	No

Note: 'Yes' indicates potential EMC problem between indicated systems; 'No' indicates no potential problem

5   BS EN 61000-6-3: 2001: *Electromagnetic compatibility (EMC). Generic standards. Emission standard for residential, commercial and light-industrial environments* (London: British Standards Institution) (2001)

6   BS EN 61000-6-2: 2001: *Electromagnetic compatibility (EMC). Generic standards. Immunity standard for industrial environments* (London: British Standards Institution) (2001)

7   BS EN 61000-6-4: 2001: *Electromagnetic compatibility (EMC). Generic standards. Emission standard for industrial environments* (London: British Standards Institution) (2001)

8   IEC 61000-2-5: 1995: *Electromagnetic compatibility (EMC); Part 2: Environment; Section 5: Classification of electromagnetic environments* (Geneva, Switz: International Electrotechnical Commission) (1995)

9   BS EN 50130-4: 1996: *Alarm systems. Electromagnetic compatibility. Product family standard: Immunity requirements for components of fire, intruder and social alarm systems* (London: British Standards Institution) (1996)

10   BS EN 50091-2: 1996: *Specification for uninterruptible power systems (UPS). EMC requirements* (London: British Standards Institution) (1996)

11   BS EN 50270: 1999: *Electromagnetic compatibility. Electrical apparatus for the detection and measurement of combustible gases, toxic gases or oxygen* (London: British Standards Institution) (1999)

12   BS EN 60730-2: *Specification for automatic electrical controls for household and similar use. Particular requirements* (22 parts) (London: British Standards Institution) (1997–2002)

13   BS EN 55022: 1998: *Information technology equipment. Radio disturbance characteristics. Limits and methods of measurement* (London: British Standards Institution) (1998)

14   BS EN 12015: 1998: *Electromagnetic compatibility. Product family standard for lifts, escalators and passenger conveyors. Emission* (London: British Standards Institution) (1998)

15   BS EN 12016: 1998: *Electromagnetic compatibility. Product family standard for lifts, escalators and passenger conveyors. Immunity* (London: British Standards Institution) (1998)

16   BS EN 55015: 2001: *Limits and methods of measurement of radio disturbance characteristics of electrical lighting and similar equipment* (London: British Standards Institution) (2001)

17   BS EN 60947: *Specification for low-voltage switchgear and control gear* (8 parts) (London: British Standards Institution) (1996–2003)

18   BS EN 50174: *Information technology. Cabling installation; Part 1: 2001: Specification and quality assurance; Part 2: 2001: Installation planning and practices inside buildings; Part 3* (draft for comment): 2002: *Installation planning and practices outside buildings* (London: British Standards Institution) (2001, 2002)

19   BS EN 50310: 2000: *Application of equipotential bonding and earthing in buildings with information technology equipment* (London: British Standards Institution) (2000)

20   BS EN 50173-1: 2002: *Information technology. Generic cabling systems. General requirements and office areas* (London: British Standards Institution) (2002)

21   BS EN 50098: *Customer premises cabling for information technology; Part 1: 1999: ISDN basic access; Part 2: 1996: 2048 kbit/s ISDN primary access and leased line network interface* (London: British Standards Institution) (1999, 1996)

22   Allen S G, Chadwick P J, Dimbylow P J, McKinlay A F, Muirhead C R, Sienkiewicz Z J and Stather J W *1998 ICNIRP Guidelines for Limiting Exposure to Time-Varying Electric, Magnetic and Electromagnetic Fields (up to 300 GHz): NRPB Advice on Aspects of Implementation in the UK* (Didcot: National Radiological Protection Board) (1998)

12   Inspection, testing, operation and maintenance

12.1   Introduction

It is a requirement of the Electricity at Work Regulations 1989[1] that any electrical system be constructed and maintained as to prevent, so far as reasonably practicable, danger.

BS 7671[2] states that:

> Every installation shall, during erection and/or on completion before being put into service be inspected and tested to verify, so far as is reasonably practicable, that the requirements of the regulations have been met.

The importance of these activities is such that they must be carried out by a competent person and the results of the inspection and test is validated against the design criteria.

Records of the inspection and test must be included in operational manuals, together with precise details of each item of equipment.

12.2   Scope

This section relates mainly to inspection, testing and maintenance of low voltage installations although reference is made to high voltage equipment where it would form part of the end-user's electrical installation.

Reference is made to the inspection and testing requirements of BS 7671[2] and IEE Guidance Note 3: *Inspection and Testing*[3], together with additional requirements of the Electricity at Work Regulations[1], Health and Safety Executive guidance and relevant British Standards.

It does not include the requirements and procedures for emergency lighting, general lighting and fire alarm systems.

12.3   Safety

Electrical testing inherently involves some degree of danger due mainly to the proximity to live parts during the testing procedure. This is also the case during inspection where the supply is available, for example, during periodic inspection of an installation.

The Electricity at Work Regulations 1989[1] must be complied with during all activities related to an electrical system. Regulation 13 requires that adequate precautions shall be taken to prevent equipment which has been made dead from becoming electrically charged during the work to be undertaken if danger may arise.

HSE guidance[4] on the Electricity at Work Regulations highlights the need for safety isolation procedures to be formalised in written instructions. Where appropriate, a 'permit-to-work' may form part of that procedure.

Where work is to be carried out on or near a live conductor, regulation 14 requires that the following criteria must be met:

(*a*)   it is unreasonable in all circumstances for it to be dead

        and

(*b*)   it is reasonable in all the circumstances for him[/her] to be at work on or near it whilst it is live

        and

(*c*)   suitable precautions (including where necessary the provision of suitable protective equipment) are taken to prevent injury.

It follows that the inspection and test activities must be carried out by a competent person and be carefully planned to ensure that no danger will arise to either persons or property.

12.4   Initial verification

12.4.1   General

BS 7671[2] clearly sets out the requirements of inspection during erection and on completion of an installation.

To be effective inspection activities must be undertaken at the correct time in the construction process. In planning the inspection and testing activities the intentions of the client, regarding witnessing, should be ascertained.

In order to carry out the work the inspector must be in possession of information regarding the installation. BS 7671 sections 311, 312, 313 require that the following are determined prior to an inspection or before an alteration or addition to the installation:

—   the maximum demand

—   the arrangement of live conductors and the type of earthing

—   the nature of the supply

—   the design criteria.

12.4.2   Initial inspection

The inspection should be carried out to confirm that the installed electrical equipment is:

— in compliance with section 511 of BS 7671[2]

— correctly selected and erected in accordance with BS 7671

— not visibly damaged or defective so as to impair safety.

The following forms a schedule of items that, where relevant, should be inspected as a minimum. For more detail on the items listed reference should be made to BS 7671 and IEE Guidance Note 3: *Inspection and Testing*[3].

(a)   Connection of conductors.

(b)   Identification of conductors.

(c)   Routing of cables in safe zones or protection against mechanical damage.

(d)   Selection of conductors for current carrying capacity and voltage drop.

(e)   Connection of single-pole devices for protection or switching in phase conductors only.

(f)   Correct connection of accessories and equipment.

(g)   Presence of fire barriers, fire seals and protection against thermal effects.

(h)   Methods of protection against electric shock by:

— protection against both indirect and direct contact

— protection against direct contact

— protection against indirect contact.

(i)   Prevention of mutual detrimental influence.

(j)   Presence of appropriate and correctly labelled devices for isolation and switching.

(k)   Presence of under voltage protective devices.

(l)   Choice and setting of protective and monitoring devices.

(m)   Labelling of protective devices, switches and terminals.

(n)   Selection of equipment and protective measures appropriate to external influences.

(o)   Adequacy of access to switchgear and equipment.

(p)   Presence of danger notices and other warning signs.

(q)   Presence of diagrams, instructions and similar information.

(r)   Erection methods.

This list is not exhaustive and should be supplemented by additional items where appropriate. Installations which are considered as special locations, as defined in BS 7671 Part 6, must also be inspected to confirm compliance.

It should be noted that inspection is not simply a visual check. Connections and fixings can only be checked by touch. Also, the smell of burning or the sound of arcing would indicate an immediate problem which should be investigated.

12.4.3   Initial testing

The testing of a new installation or an addition to an existing installation must be carried out in a prescribed sequence and the results of those tests compared with relevant pass criteria.

12.4.3.1   Sequence of tests

Before energisation

(a)   *Continuity of protective conductors*: all protective conductors including earthing conductors, circuit protective conductors, main and supplementary bonding conductors should be tested to verify connection and continuity. Wherever possible parallel earth paths should be eliminated from the test, however this is not always possible or indeed practical. Where parallel earth paths cannot be disconnected the test should be supplemented with inspection to confirm the integrity of the protective conductors.

(b)   *Continuity of ring final circuit conductors*: the continuity of each conductor of a ring final circuit should be verified to ensure the ring is complete and is not interconnected.

(c)   *Insulation resistance*: the insulation resistance between all conductors, between conductors and earth, and of accessories should be measured and verified in accordance with BS 7671[2] Table 71A (see Table 12.1).

Care must be taken whilst performing this test to disconnect all current using equipment; where this is not possible a test to earth with phase and neutral strapped together should be made.

The test must be carried out with all switches and/or devices in the closed position to ensure all parts of the circuit are tested.

(d)   *Polarity*: the polarity of every circuit should be verified to ensure correct connection of conductors and that all single pole switches or protective devices are in the phase conductor only.

For ring final circuits the polarity is proved during the continuity of ring final circuit conductors.

For radial circuits and lighting circuits the procedure for testing continuity of protective conductors, as described in IEE Guidance

Table 12.1 Minimum values of insulation resistance

Circuit nominal voltage / V	Test voltage (DC)	Minimum insulation resistance / MΩ
Separated extra-low voltage (SELV) and protective extra-low voltage (PELV)	250	0.25
Up to and including 500 V with the exception of the above systems	500	0.5
Above 500 V	1000	1.0

Note 3[3], would effectively prove polarity. Providing all devices are operated during the test there would be no need to retest polarity prior to energisation.

(e) *Earth electrode resistance*: where applicable it will be necessary to measure the resistance of the earth electrode to the general mass of earth to ensure the earthing arrangement satisfies the requirements of both BS 7671 and the design parameters for the particular installation.

It should be noted that any test requiring the disconnection of the electrode must be performed with the supply to the installation isolated.

Where appropriate:

(f) *Site applied insulation*: where insulation is applied during erection, not including the assembly of type tested switchgear on site, a test should be performed between the live conductors and the external surfaces of the insulation.

Test voltages and duration should be in accordance with the appropriate British Standard for similar factory built equipment.

The test involves the application of high voltage and should therefore be strictly controlled by a suitably authorised and competent person.

(g) *Protection by separation of circuits*: where protection is afforded by separation of circuits utilising either separated extra-low voltage (SELV) or protective extra-low voltage (PELV) the separation should be verified between the primary and secondary live conductors and, in addition, for SELV circuits between the secondary live conductors and the protective conductors of any adjacent higher voltage system. Test voltage will be in accordance with BS 7671[2] Table 71A.

(h) *Protection by barriers or enclosures provided during erection*: where protection against direct contact is afforded by a barrier or enclosure is should be verified that the degree of protection meets IP 2X or IP XXB and IP 4X on any accessible horizontal surface. (Degrees of protection afforded by enclosures and defined in BS EN 60529[5], see section 7.3.2).

(i) *Insulation of non-conducting floor and walls*: where protection against indirect contact is provided by a non-conducting location it should be verified that under normal circumstances there can be no simultaneous contact between exposed conductive parts and extraneous conductive parts and that the resistance of insulating floors and walls to the main protective conductor is above 50 kΩ, (for a system not exceeding 500 V) where the resistance is less than that specified the wall on floor should be deemed to be an extraneous conductive part.

After energisation

(j) *Recheck polarity*: polarity checks should be carried out once the system has been energised using an approved test lamp in accordance with HSE GS 38[6].

(k) *Prospective fault current*: the prospective fault current should be measured at the origin of an installation and at all other relevant points.

The test should include both prospective short circuit current and prospective earth fault current, the greater of the two being compared to the rating of the protective device.

(l) *Earth fault loop impedance*: the earth fault loop impedance should be measured at both the origin of the installation (external earth fault loop impedance Z_e) and at the furthest point of each distribution circuit (maximum earth fault loop impedance Z_s).

When measuring the external earth fault loop impedance of an installation the earthing conductor must be disconnected from the main earthing terminal, and on any other bonding conductors and circuit protective conductors, to ensure that false readings due to parallel paths are not obtained. For this reason it is imperative that the installation is isolated from the supply during the test.

On completion of the test the earthing conductors must be reconnected before energisation of the installation. Alternatively Z_e may be determined by enquiry or by calculation, though with either of these two methods a measurement should be taken to ensure a connection with earth exists.

The measurement of maximum earth fault loop impedance (Z_s) should be taken at the remote end of all final circuits, in the case of ring final circuits at the mid-point. This will necessitate a test at each socket outlet to determine the mid-point.

Before carrying out the test, it must be confirmed that all earthing conductors, circuit protective conductors and bonding conductors are in place.

As the test instrument used for this procedure passes a test current in the order of 23 A through the phase earth loop, it may not be possible to perform the test, for example, in circuits protected by low rated MCBs or where there is a risk of damage to electronic controls. In this case it is possible to calculate the maximum earth fault loop impedance using $(R_1 + R_2)$ values obtained whilst testing the circuit protective conductors and adding the test results obtained for other parts of the circuit, i.e:

$$Z_s = Z_e + (R_1 + R_2) \qquad (12.1)$$

where Z_s is the maximum earth fault loop impedance (Ω), Z_e is the external earth fault loop impedance (this may also be the reading obtained at the distribution board or control panel supplying the final circuit under test) (Ω), R_1 is the resistance of the phase conductor (Ω) and R_2 is the resistance of the circuit protective conductor (Ω).

It should be noted that the measured Z_s value would include all parallel earth paths and this should be taken into account when making a comparison to measured $(R_1 + R_2)$ value.

Where circuits are protected by RCDs, instruments are available which, by either utilising a reduced test current or by saturating the core of the RCD, allow a test to be made without operating the device.

Where one of these instruments is not available, the method using $(R_1 + R_2)$ values should be used. At no time should the device be bypassed during the test.

(m) *Functional test of devices including residual current operated devices*: a series of tests should be performed on residual current devices to test their effectiveness after installation. The tests are in addition to the proving of the integral test button, which must be checked after the test sequence.

The test sequence is as follows:

(1) At a test current of 50% of the rated tripping current of the RCD, the device should not open.

(2) At a test current of 100% of the rated tripping current of the RCD, the device should operate in less than 200 ms for general purpose RCDs to BS 4293[7], and less than 300 ms for general purpose RCDs to BS EN 61008[8] or RCBOs to BS EN 61009[9].

(3) Where the device provides supplementary protection against direct contact, an additional test should be carried out at 500% of the rated tripping current. The device should operate in less than 40 ms.

Where the device incorporates a time delay, the manufacturer's instructions should be consulted.

(n) *Functional test of switchgear and control gear*: all site built assemblies must be tested to prove that they are installed in accordance with regulations and in accordance with manufacturers instructions. With many assemblies such tests may be carried out prior to dispatch therefore eliminating difficult and often expensive testing on-site.

The following forms a list of items that, where applicable, should be proved during manufacturers' acceptance testing:

(a) HV switchgear:

— Physical inspection to ensure compliance with specification and to confirm the general arrangement is as per the design. It should also be confirmed that the equipment is suitable for operation under the end user's safety rules.

— Function test to ensure correct operation of equipment.

— Insulation resistance and pressure test at appropriate voltages, to approved standards.

— Primary injection of overcurrent, earth fault protection, to prove sensitivity and selectivity and CT ratio.

— Secondary injection of protection relays in accordance with manufacturer's commissioning guides to prove correct operation and to confirm the operation of the trip circuit.

— Appropriate injection of other devices.

(b) LV switchgear:

— Physical inspection to ensure compliance with the specification and to confirm the general arrangement is as per the design.

— Primary injection of restricted earth fault protection to prove sensitivity when current is passed through one CT and selectivity when current is passed through two CTs .

— Primary and secondary injection of overcurrent protective devices in accordance with manufacturer's commissioning guides to prove correct operation. Where the device and all its components are contained within one unit the device manufacturer's test certificates would be acceptable. However, where components, e.g. current transformers, are mounted and wired by the panel manufacturer then tests must be carried out to verify the components are correctly installed.

— Function test to prove correct operation of equipment including both mechanical and electrical interlocks.

— Primary insulation resistance tests and voltage withstand tests in accordance with appropriate British Standards.

— Confirmation of neutral earthing arrangements.

Following erection on site the following tests must be carried out:

(a) HV switchgear:

— Physical inspection to ensure erection to manufacturer's recommendations.

— Busbar joint integrity test by either torque or resistance method.

— Insulation resistance, pressure test, to manufacturer's recommendations.

— Secondary injection test of protection relays to prove operation at service settings.

— Primary injection of unit protection schemes to prove sensitivity and selectivity.

— Overall function check to prove correct operation of equipment and controls, and intertripping.

(b) HV cables:

— Physical inspection to ensure that the cable installation is complete and in accordance with the design. It should also be confirmed that methods of cleating and bending radii of cables are compliant with manufacturer's guidance.

— Terminations of cables should be inspected to confirm installation as per the manufacturer's instruction, and be checked for correct separation of conductors within dry termination boxes. The minimum distance between conductors and between conductors and earth for a paper insulated cable to BS 6480[10] is 10 mm at the 'crutch' of the termination increasing immediately to 20 mm. For a XLPE insulated screened cable to BS 6622[11] a minimum distance of 15 mm between conductors and between conductors and earth should be met at the end of the screen increasing to 20 mm at the end of the stress relieving tubing.

— A test should be carried out to prove continuity and polarity.

— An insulation resistance test should be carried out at 1000 V.

— The cable should be pressure tested in accordance with the appropriate British Standard. Table 12.2 shows test voltages for new cables. These test voltages should be applied for a period of 15 minutes.

Where the installation comprises cables that have been in service, then a reduced test voltage should be applied. The test voltage should be agreed with both the cable manufacturer and the client.

— Following a pressure test, the insulation test should be repeated.

(c) LV switchgear:

— Physical inspection to ensure erection to manufacturer's recommendations.

— Busbar joint integrity test by either torque or resistance method.

— Insulation resistance test in accordance with BS 7671[2].

— With service setting applied protection relays should be secondary injected to prove operation. Where the device or its wiring will not be affected by transportation the tests performed at the manufacturer's works would suffice.

— An overall function check to prove correct operation of equipment, controls, intertripping and interlocks.

— Neutral earthing arrangements to be confirmed and recorded.

— Following energisation the phase rotation and voltage should be checked and recorded.

(d) Transformers:

— Physical inspection to ensure installation to manufacturers recommendations.

— Insulation resistance: pressure test to manufacturer's recommendations.

— Over-temperature control to be proved to operate correctly.

— Buchholz device operation to be proved to operate correctly, where applicable.

12.4.4 Verification of test results

It is the responsibility of the person carrying out the inspection and test activities to verify that the results comply with relevant pass/fail criteria. It is therefore essential that the designer provides the information to the person carrying out the construction and, where different, the inspection and testing. Where this is not available, reference should be made to appropriate tables in BS 7671[2] and/or other relevant documents.

In relation to the measurement of earth fault loop impedance IEE Guidance Note 3[3] details four methods of verifying test results, these are as follows:

(a) Table 41B or Table 41D of BS 7671, after correction for temperature.

Where conductors are tested at a temperature which is different to its normal operating temperature, adjustment should be made to the readings. The formula for this adjustment is:

$$Z_e + F(R_1 + R_2)_{test} < Z_s \qquad (12.2)$$

where Z_e is the external earth fault loop impedance (Ω), F is the conductor temperature resistance factor (see IEE Guidance Note 3[3], Appendix 1, Table 1C), $(R_1 + R_2)_{test}$ is the measured value of $(R_1 + R_2)$ at the ambient temperature during the test (Ω) and Z_s is the maximum earth fault loop impedance (see BS 7671[2], Table 41B1, 41B2, 41D) (Ω).

(b) Earth fault loop impedance figures provided by the designer.

IEE Guidance Note 3[3] gives the following formula, which may be used by designers to give maximum Z_s values:

$$Z_{test} \leq (Z_s / F) \qquad (12.3)$$

where Z_{test} is the earth fault loop impedance obtained when testing an installation at an ambient temperature of 20 °C (Ω), Z_s is the maximum earth fault loop impedance (see BS 7671[2] Table 41B1, 41B2, 41D) (Ω) and F is the conductor temperature resistance factor (see IEE Guidance Note 3[3], Appendix 1, Table 1C).

(c) Rule of thumb.

The measured value should not exceed 75% of the figures given in Table 41B1, 41B2, 41D of BS 7671[2].

This provides a relatively quick method of verifying that the circuit will operate within the appropriate disconnection time. However, should

Table 12.2 Test voltages for cables up to 33 kV

Cable type	Test voltage phase-to-phase	Test voltage phase-to-earth
Paper insulated to BS 6480[ref]	34 kV (DC)	25 kV (DC)
XLPE insulated to BS 6622[ref]	Not applicable	25 kV (DC)

the test value be greater than the 75% value, further investigation is required by either of the other methods mentioned or by discussion with the designer.

(d) For standard PVC circuits, values are given in Appendix 2 of IEE Guidance Note 3[3].

A comparison can be made of Z_s test values against the tables; as these figures are based on tests being carried out at an ambient temperature of 10–20 °C no correction need be carried out under normal test conditions.

12.5 Periodic inspection and test

12.5.1 General

All electrical installations will deteriorate to some extent throughout their expected lifetime. This could be due to damage or wear and tear, but could also be due to excessive loading or adverse environmental influences.

For this reason it is necessary to carry out periodic inspection supplemented by tests at intervals dependant on the type of installation.

The requirement for electrical installations to be maintained in a safe condition dictates that periodic inspection and testing be carried out. In the instance of licensed properties, periodic inspection is essential in the renewal of a licence.

Prior to an inspection being carried out it is necessary to obtain relevant diagrams, schedules and information regarding previous inspections, and any alterations or additions. Should these not be available some survey work would be required in order to ensure that the inspection and testing can be carried out in a safe and effective manner.

12.5.2 Periodic inspection

Inspection should take into account the following with regard to the electrical installation and all connected electrical equipment:

— safety
— wear and tear
— corrosion etc by external influence
— damage
— excessive loading/over heating
— age of the installation
— suitability of installed equipment
— function of the installation.

Individual inspections should include items as follows:

— a sample of joints and connections
— identification and condition of conductors

— condition of flexible cables and cords
— internal inspection of a sample of switching devices
— presence and integrity of fire barriers where reasonably practicable
— means of protection against direct contact with live conductors
— means of protection against indirect contact
— presence, identification, condition and accessibility of protective devices and switching devices
— condition and integrity of enclosures and mechanical protection
— correct labelling of installations with regard to next inspection, earthing, voltages and presence of residual current devices
— any changes in either the external influences which may effect the installations or alterations and additions that are not appropriate to the particular external influences present.

12.5.3 Periodic testing

As stated in section 12.5.1, periodic inspection should be supplemented by testing. The level of testing may not be the same as that for an initial test. However, should a sample test indicate either deterioration or failure, then sampling must be increased.

It should be noted that periodic testing involves a potentially increased risk to safety and therefore the use of competent persons with careful planning and preparation is essential.

12.5.4 Test procedures

The methods for carrying out periodic tests are the same as those for initial verification. However, a different sequence of test should be adopted to reflect the need to confirm the integrity of the earthing prior to carrying out further tests.

The following sequence should be used:

(1) Continuity or protective conductors including earthing and bonding conductors: care must be taken whilst carrying out these tests as a dangerous potential can exist between CPC and the main earthing terminal when the CPC is disconnected particularly where the CPC provides functional earthing. For this reason no earthing conductors should be disconnected with an installation energised.

(2) Polarity tests should be carried out, however, where the testing of the CPC utilises the $(R_1 + R_2)$ method which incorporates polarity testing as described in IEE Guidance Note 3[3]. No further test need be performed.

(3) Earth fault loop impedance tests must be carried out initially at the point of the incoming supply and then at the remote end of each radial circuit and at every socket outlet.

(4) An insulation resistance test should be carried out between conductors and between conductors and earth. All equipment susceptible to damage by this test should be disconnected. Where this is not possible, only a test to earth can be performed. This should be carried out with all phase and neutral conductors strapped together.

(5) All devices used for switching and isolation should be proved to function correctly.

(6) All residual current devices should be proved to function correctly by simulating earth fault conditions and by the operation of any integral test buttons.

(7) Overcurrent protective devices should be proved to function manually.

(8) Where there is evidence of alteration, it would also be required that the continuity of ring final circuit conductors is proved.

12.6 Certification

12.6.1 General

It is a requirement of BS 7671[2] that, on completion of a new installation or where additions or alterations have been made to an existing installation, certification must be produced. It is also a requirement that, where a periodic inspection of an installation is completed, a periodic inspection report is issued. All certification must be signed by a competent person or persons.

Model forms can be found in BS 7671 Appendix 6.

12.6.1.1 Electrical installation certificate

For a new installation or for an addition or alteration which includes the provision of a new circuit, an electrical installation certificate must be issued upon completion of the works.

The certificate should include the following:

— extent of the installation covered by the certificate

— schedules of inspection and testing

— schedules of test results and circuits details

— details of persons responsible for design, construction, and inspection and test.

In the case of alterations or additions, any defects found on the existing installation should be recorded on the certificate. (Defects which would have an effect on the safety of the new work must be rectified before the installation is put back in to service.)

12.6.1.2 Minor electrical installation works certificate

Where an addition or alteration does not include the provision of a new circuit a minor electrical installation works certificate may be issued.

The model for this certificate given in BS 7671 Appendix 6 is a single page certificate, and is therefore appropriate for this type of work.

It should be noted that it is not acceptable to use this certificate for additions or alterations to more than one circuit, nor for replacement of switchgear not on a 'like for like' basis or where the switchgear incorporates protective devices for more than one circuit.

12.6.1.3 Periodic inspection and testing

Following a periodic inspection, a periodic inspection Report must be issued and should include the following:

— the extent of the installation covered by the report

— agreed limitations, if any, on the inspection

— the purpose for which the report has been carried out (whether this be following a fire or flood, for licensing application or at the end of a recommended period)

— observations and recommendations for actions to be taken; any recommendations should be categorised using a code number as follows:

    (1) Requires urgent attention.

    (2) Requires improvement.

    (3) Requires further investigation.

    (4) Does not comply with BS 7671 (this does not imply that the electrical installation inspected is unsafe).

— a summary of the inspection detailing the condition of the installation with regard to safety

— a schedule of inspection, testing and test results.

12.7 Operation and maintenance procedures

12.7.1 Operation and maintenance manuals

An essential part of any building services contract is the provision of an operation and maintenance manual to the client or end user upon completion of the contract.

This document should form part of the effective operation of the system and as such should be carefully prepared to include a comprehensive source of reference of the installed equipment.

The information within the manual should also provide for the safety of the end user and as a basis for the training of personnel in the operation of the installation.

One of the most important uses of the manual is as a reference base for use during modification or additions to an installation and during periodic inspections. It is therefore necessary that following all modifications to the installation, the operation and maintenance (O&M) manual is updated to reflect the changes made.

12.7.2 Contents

The O&M manual should contain the following information:

— the purpose and use of the installation

— the technical specification of the installation

— 'as fitted' drawings

— instruction in the operation of the installation

— recommended maintenance procedures

— recommended maintenance schedules

— a list of recommended spares

— instructions in the modification of plant or system where permitted by the system designer or manufacturer; space should also be provided for records of the modifications

— instructions for the disposal of equipment with regard to the safety of persons and the environment

— details of all manufacturers and suppliers of equipment together with details of the parties responsible for design, construction and installation.

12.7.3 Recommended maintenance procedures

In order to ensure that systems remain in a safe condition, it is necessary to carry out periodic inspection see section 12.5. However this is not sufficient on its own as damage or deterioration can occur on a daily basis. It is therefore necessary that systems be subjected to a maintenance routine appropriate to the use and type of the installation.

Maintenance can be broken down in to three categories:

— *Breakdown maintenance*: replacement or repair of equipment as and when it breaks down; clearly this simplest approach to maintenance is only appropriate to certain types of equipment and installation.

— *Preventative maintenance*: planned maintenance of equipment to a prepared schedule to avoid the breakdown of equipment.

— *Condition monitored maintenance*: this requires that equipment is monitored during its operation for signs of deterioration, e.g. temperature readings, from which maintenance can be carried out to prevent breakdown.

Recommended schedules for maintenance procedures are detailed in HVCA Standard Maintenance Specification for Mechanical Services in Buildings[12].

12.8 Test instruments

12.8.1 Safety

To comply with the requirements of the Electricity at Work Regulations 1989[1], it is essential that test equipment be constructed, maintained and used in such a way as to prevent danger.

Accidents have occurred due to unsuitable test probes, excessive currents being drawn through test probes when instruments are incorrectly used, and where makeshift test equipment i.e. test lamps are used.

In order to reduce the risk involved in the use of test equipment, a number of documents have been produced, including the following:

— HSE GS 38: *Electrical test equipment for use by electricians*[6]

— BS EN 61010: *Safety requirements for electrical equipment for measurement, control and laboratory use*[13]

— BS EN 61557: *Electrical safety in low voltage distribution systems up to 1000 V AC and 1500 V DC*[14].

HSE GS 38 requires that test leads and probes have the following safeguards:

— finger barriers or a shape to prevent accidental contact with the conductors under test

— insulated to leave a maximum of 4 mm of exposed conductors at the tip of the test probe

— fused or current limiting resistor and a fuse

— adequately coloured to distinguish between each lead/probe

— leads of an adequate length to allow the test to be performed safely

— should leads become detached during a test, no conductive parts other than the tip of the probe can become accessible.

12.8.2 Accuracy

It is necessary that the accuracy and consistency of test equipment is maintained by calibration at intervals recommended by the instrument manufacturer, and certificates of the calibration be produced. It follows that in maintaining the calibration status of equipment all items of equipment must be identifiable by a unique serial number. This also provides for traceability of instruments from test records.

The performance criteria of test instruments are detailed in BS EN 61557[14] and are also summarised in IEE Guidance Note 3[3].

References

1   The Electricity at Work Regulations 1989 Statutory Instruments 1989 No. 635 (London: Her Majesty's Stationery Office) (1989)

2   BS 7671: 2001: *Requirements for electrical installations. IEE Wiring Regulations. Sixteenth edition* (London: British Standards Institution) (2001)

3   *Inspection and testing* IEE Guidance Note 3 (London: Institution of Electrical Engineers) (2002)

4   *Memorandum of guidance on the Electricity at Work Regulations 1989* Health and safety booklet HS(R)25 (London: Health and Safety Executive) (1998)

5   BS EN 60529: 1992: *Specification for degrees of protection provided by enclosures (IP code)* (London: British Standards Institution) (1992)

6   *Electrical test equipment for use by electricians* General Series Guidance Note GS38 (London: Health and Safety Executive) (1995)

7   BS 4293: 1983: *Specification for residual current-operated circuit-breakers* (London: British Standards Institution) (1983)

8   BS EN 61008: *Specification for residual current operated circuit-breakers without integral overcurrent protection for household and similar uses (RCCBs)*; Part 1: 1995: *General rules*; Part 2: 1995: *Applicability of the general rules to RCCBs functionally independent of line voltage* (London: British Standards Institution) (1995)

9   BS EN 61009: *Specification for residual current operated circuit-breakers with integral overcurrent protection for household and similar uses (RCBOs)*; Part 1: 1995: *General rules*; Part 2: 1995: *Applicability of the general rules to RCBOs functionally independent of line voltage* (London: British Standards Institution) (1995)

10   BS 6480: 1988: *Specification for impregnated paper-insulated lead or lead alloy sheathed electric cables of rated voltages up to and including 33000 V* (London: British Standards Institution) (1988)

11   BS 6622: 1999: *Specification for cables with extruded cross-linked polyethylene or ethylene propylene rubber insulation for rated voltages from 3.8/6.6 kV up to 19/33 kV* (London: British Standards Institution) (1988)

12   *Standard Maintenance Specification for Mechanical Services in Buildings* HVCA SFG 20/M1 (CD-ROM) (London: Heating and Ventilating Contractors' Association) (2004)

13   BS EN 61010: *Safety requirements for electrical equipment for measurement, control and laboratory use* (19 parts) (London: British Standards Institution) (1993–2003)

14   BS EN 61557: *Electrical safety in low voltage distribution systems up to 1000 V AC and 1500 V DC* (10 parts) (London: British Standards Institution) (1997–2001)

Index

GUILDFORD **college**

7 DAY
BOOK

Learning Resource Centre

Please return on or before the last date shown.
No further issues or renewals if any items are overdue.

Skillingsgt?
2005
(Dec 18,)
Also on CIS
with correction
+ corrigenda
Jan '06

Class: 696 CHA

Title: CIBSE GUIDE K: ELECTRICITY IN BUILDINGS

Author: CIBSE